AF577282

Fragile Wings & Gentle Giants

Fragile Wings &

Gentle Giants

by Harold Salut

Blackbird Press, Dubuque, Iowa

Copyright © 1985 by Harold Salut.
All rights reserved.

Published by Blackbird Press
Dubuque, Iowa 52001

No part of this manuscript may be reproduced
in any manner without prior
written permission from the publisher.

First edition, 1985
Printed in the United States of America
ISBN 0-933473-01-X

This book is dedicated to
all the 'gentle giants'
named herein.
Men for whom
the beauty and excitement
of flight
was an
irresistable call.

AN AIRMAN'S HYMN

When the last flight is over
And the happy landing past,
And the altimeter tells me
The big crash has come at last,
I'll swing her nose up to 'high noon'
And give my crate the gun ,
And open her up and let her zoom
For that airport on the sun.

And the great God of flying men
Will smile at me sort of slow
As I stow my crate in the hangar
On the field where flyers go.
And then I'll look upon His face,
The Almighty Flying Boss,
Whose wingspread covers the sky
From Orion to the Cross

FORWARD

Many people keep diaries, and the more ambitious might conjur up ledgers of names, dates and places. Both are commendable endeavors . . . but I decided to write a book.

The memories of special friends and events have always been close to me, and in a sense, created or greatly influenced my own destiny.

But the memories grow dim with passing time, events become distorted and reality fades into that netherland of myth and legend.

So I have written this book intent on rekindling the memories of long departed friends. Their triumphs and their failures might add another small facet to the history of man's quest for the unknown.

INTRODUCTION

After World War I, the embryo aviation industry slipped into the doldrums of a postwar economy.

In May 1918, the U.S. Aerial Mail Service, under the Postoffice Department, was inaugurated by President Wilson and was the first commercial venture of the airplane.

Smaller commercial ventures were carried on by the "barnstormers". These were individuals or small groups of flyers who crossed the country making what money they could by giving flying shows and taking passengers for 'aeroplane' rides. They moved across the countryside, stopping at each small town, drumming up what business they could, sleeping under the wings of their planes at night and moving on to the next town the following day.

In 1925 President Coolidge signed the Kelly bill authorizing contract airmail to private bidders, and the following year, the Air Commerce Act of 1926 authorized the development of commercial aviation and establishment of air routes.

Charles Lindbergh's historic solo flight from New York to Paris on May 21, 1927 captivated the world and gave the aviation industry a much needed boost. Passenger flying was encouraged and mail planes gravitated from single engine transports to multi engine liners.

Barnstormers were still enjoying their heydey during this era and were responsible for introducing the citizens throughout the land to their first glimpse, and frequently, their first ride in an airplane.

This story tells some of the experiences of one young man had during the summer and winter of 1932 while flying for one of these barnstorming aerial circuses.

ACKNOWLEDGEMENTS

My thanks to the many friends who encouraged me to put these memories together. names slip away with time but friends are never forgotten.

A special thanks to Bill Rhode, Vince Doyle, Cole Palen, George Hardie, Carl 'Slim' Hennicke, and the Boeing Company for the use of the photos in the book. These dedicated people help keep aviation history alive and I'm grateful to be associated with them.

A special thanks also to Bob Taylor, president and founder of the Antique Airplane Association; and to Paul Poberezney, president and founder of the Experimental Aircraft Association. Both of these men have contributed greatly to the preservation of aviation history. Their museums at Blakesburg, Iowa and Oshkosh, Wisconsin allow present and future generations to peer through a window into the aeronautical past and sense that same excitement felt by that lusty breed of aviators who are legends of a bygone era.

And finally a grateful thanks to Paula Hoshall for her perserverance in typing the first drafts of this book. It was truly a heroic effort to decipher my illegible scrawl.

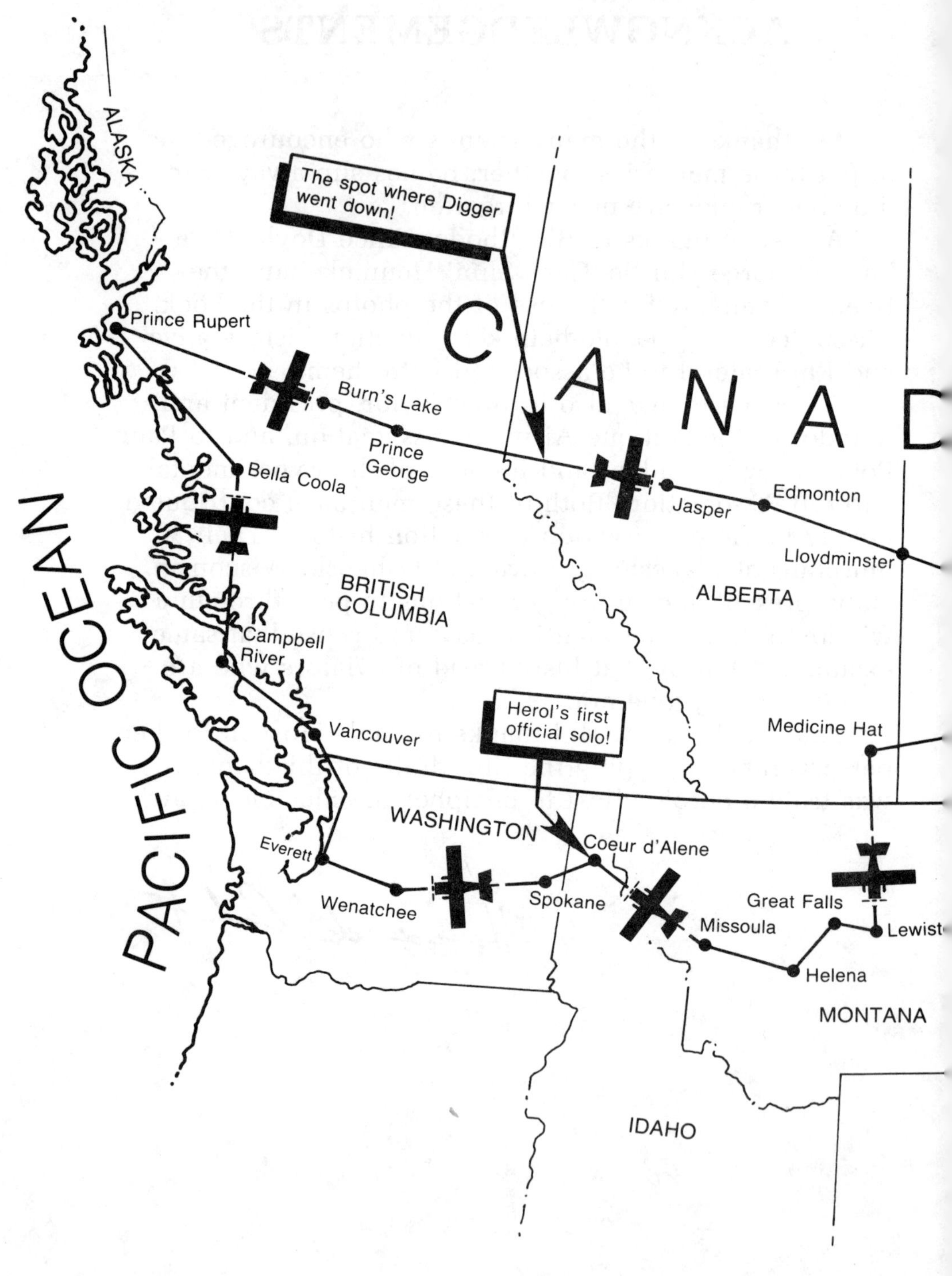

The spot where Digger went down!
Herol's first official solo!
ALASKA
CANAD
PACIFIC OCEAN
Prince Rupert
Burn's Lake
Prince George
Bella Coola
Jasper
Edmonton
Lloydminster
BRITISH COLUMBIA
ALBERTA
Campbell River
Vancouver
Medicine Hat
WASHINGTON
Everett
Coeur d'Alene
Wenatchee
Spokane
Great Falls
Missoula
Lewist
Helena
MONTANA
IDAHO

A MAP of southwestern Canada and northwestern United States showing the route author Harold 'Herol' Salut took on his odyssey with the Nord Air Flying Circus back in 1932.

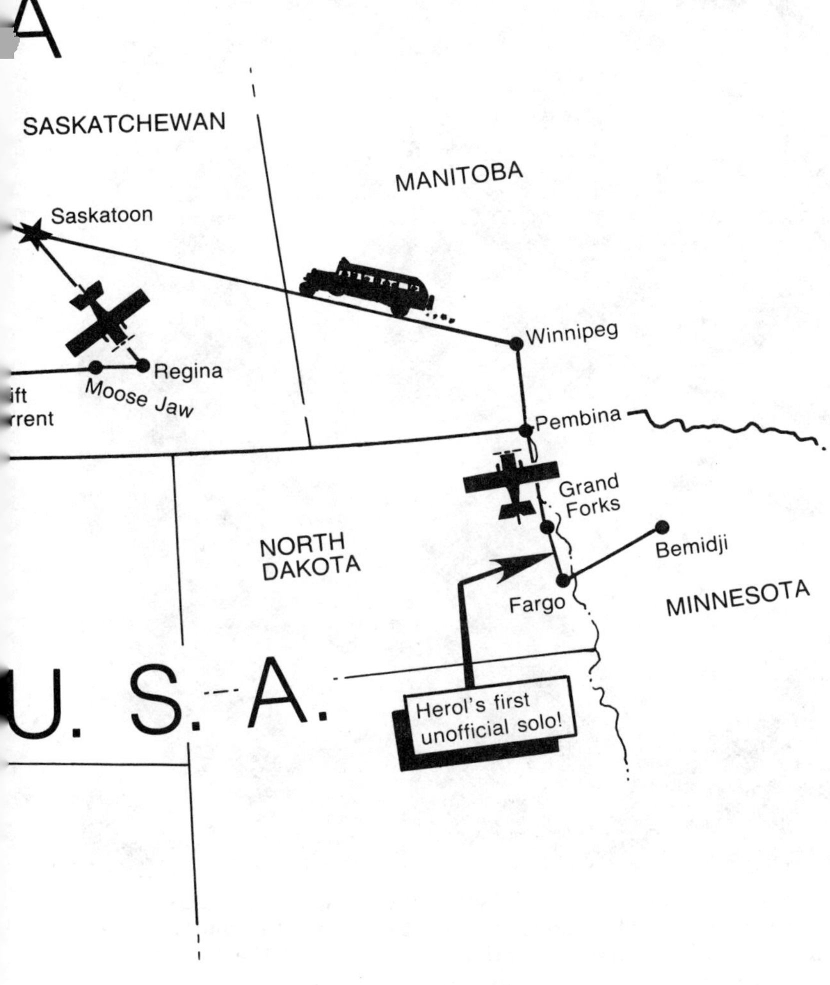

Author Harold Salut eases down into the cockpit of a Vought 03U-3 biplane. This picture was taken about seven years after the events in this story took place, by which time the author was flying for the U.S. Navy aboard the aircraft carrier U.S.S. Lexington. (Author's collection)

1 THE BARNSTORMER

It was mid October, 1919. The great World War had been over less than a year. America had made the world safe for democracy and optimism was spreading across the nation. Factories were running at peak capacity turning out civilian products that had been neglected during the recent conflict. Doughboys were settling down to the good life with their wartime brides. Henry Ford was turning out record numbers of 'Tin Lizzies', and everyone was 'goin' like sixty.'

From this horn of plenty emerged a small band of men who lived on the fringe of society. They were the dreamers, the loners, and drifters who couldn't accept the transition back to the dull routine of everyday life after a taste of the exciting and hectic war years.

A special breed emerged from this group of outcasts—the 'barnstormer'. He was an aerial vagabond who danced in the skies with a stick and wire contraption that was covered with cotton cloth. It was called an aeroplane, flying machine, airy plane, or a double winger, depending on the expertise of the lucky person who might see one.

The 'barnstormer' haunted the county fairs, rural celebrations, holiday socials, most any place where the country bumpkin or city slicker would part with his money. His only requirement was a nearby cow pasture, stubble field, or clearing where he could attract an audience.

His bag of tricks unfolded with a sequence of derring-do, foolhardy risks, plus noise and ballyhoo. And when the earth-bound spectators were at their peak of excitement, he landed in front of them and 'passed the hat' for whatever nickels and dimes they might offer. One or two of the braver onlookers could usually be induced into taking their first airplane ride. Once this happened, the money rolled in, one dollar, five dollars a head, whatever the traffic would bear.

These aerial vagabonds who danced to the tune of nickels and dimes inspired every emotion from contempt to sheer adoration by the lesser mortals who marveled at their devil-may-care existence. He was the aerial acrobat, hero, or damned fool, depending on the point of view. In reality, he was a dusty booted, leather clad, nameless aviator—a portent of the air age to come.

'Muggy' Pederson was a barnstormer. Muggy was a loner, possibly a dreamer, and surely a drifter. He was an old man beyond his years, yet he was only thirty six. He'd lived the past thirteen years in the air, a remarkable accomplishment since the life expectancy of a pilot was usually measured in months. Flying was a way of life to him long before the crop of wartime trained flyers ever saw a plane. He'd competed in exhibition flights with the legendary Wright brothers and Glen Curtiss. His name was synonymous with the great and the near great. He was a legend in the flying fraternity and was 'barnstorming' long before the word was coined. Muggy was often heard of but seldom seen. The younger generation of flyers considered him a figment of countless stories that were enhanced with each telling. Yet, his legend persisted.

Muggy had been wandering across the length and breadth of the southwestern United States for several months. Both he and his fragile craft were tired and in need of a long rest. Their odyssey had taken them along the length of the Mexican border to the bayou country of Louisiana, then northward through the Mississippi delta country. And now, after many weeks, they were nearing the land of his birth, the northern Minnesota lake country.

The Avro biplane was Muggy's trusted friend, constant companion, and his most prized possession. The Avro was a Canadian built, war surplus craft that he'd purchased shortly after World War I was over. He'd instructed young fledglings during the war in similar craft while employed as a civilian instructor by the U.S. Signal Corps. Even then, he was considered too old to fly in uniform, yet he flew as a civilian and was still flying.

Muggy and the Avro flew an erratic course through billowing cumulus clouds. They appeared as a dainty insect against the white backdrop of churning clouds and patches of blue sky. Sunlight radiated between the columns of white vapor, and cast an irridescent sheen that filtered through the weathered, cotton fabric, that covered the wings and fuselage, a silhouette of wooden ribs and wires. The total picture was one of fragile grandeur, the flimsy craft and its hardy master pitted against the unlimited whims of this sea of air.

Several hundred feet below, a patchwork of greens, browns and yellows scrambled in all directions to fade into the distant horizon. The numerous lakes sparkling with icy water mirrored puffs of cotton on their surfaces as clouds drifted from one shore to the next. The autumn days were rapidly adding new colors to the panorama below.

It was unseasonably cold, and Muggy shuddered at the thought of another lonely vigil under the Avro's wings in an unknown pasture. As if by instinct, the plane also shuddered. A staccato bark and a whirlwind of oily smoke blanketed the cockpit. Cold sweat oozed behind Muggy's goggles and dripped into his eyes as he frantically adjusted the fuel/air ratio of the cantankerous LeRhone rotary engine. Very slowly, the crisis passed and the valiant Avro struggled to keep flying. The numbing cold crept through Muggy's bones and a vague uneasiness plagued him.

"Guess the old gal is as cold as I am," he mumbled between chattering teeth.

Vibration telegraphed across the engine cowling and disappeared in the wings. Moments later a loud scream erupted overhead, focusing Muggy's attention on the cabane strut where a wind driven fuel pump shrieked and spun its tiny propellor through an erratic arc.

"Bearing must have seized, guess I shoulda used lighter grease for the cold weather," he mouthed through frosty lips. He wiped the grime off his goggles with the back of his leather mitten and felt the clammy sweat on his palms inside the woolen liners.

The sun was sinking lower, and soon the stars would reach out and pull down the night sky to cover the day. They'd have to land in a few miutes or risk a night crackup. The terrain began to look familiar, the pattern of lakes and logging camps were boyhood haunts of long ago. For fleeting moments, he felt a sense of security by ignoring the ailing engine that shuddered in protest yet continued to run.

A sharp bark interrupted his thoughts. Dense oily smoke poured from the exhaust ports, then silence, except for the gentle murmur of rigging wires between the wings that whispered to the wind.

The ailing craft inclined its head toward the deepening shadows as frostbitten fingers fumbled in the cockpit to find a fuel setting that might revive the dieing engine.

Bang! It caught hold with a harsh growl and a blanket of smoke. They had a new lease on life for the time being. The Avro raised its head a bit and Muggy smiled as he wiped the perspiration from the back of his neck. He wanted to say a prayer of thanks but didn't know how, so he blessed Lady Luck for one more chance.

Lake Itasca passed slowly beneath the wings as they limped along their uncertain path. He was sure of where they were at now. The Avro had never been here before, but she trusted her companion and master and continued her weary way. Dense timber forests attempting to swallow the shoreline, cast lengthening shadows across the water, and the feeble sunlight dancing on the center of the lake caught the reflection of the plane as it bobbed like a small boat from one ripple to another.

A dark mirrored surface emerged from the horizon, and if his guess was right, it would be Lake Plantaganet. The terrain west of the lake was bathed in the glare of the setting sun, hiding a small clearing in the woods about a mile from the shoreline. Muggy forced himself to relax. It was time to think straight or they'd be in trouble before long.

The weary Avro sank a bit deeper into the evening haze as they neared the shores of Lake Plantaganet. Muggy squinted through oil spattered goggles at the shoreline and the surrounding terrain; hunting for that safe haven they needed to find before nightfall. As he peered through the twilight shadows he spotted two figures on the shoreline. An old man and a small boy were silhouetted against the sand. They'd been waving furiously long before he'd ever seen them. As he approached along the shore, the old man danced a jig and the small boy ran in circles, continuing to wave furiously.

When Muggy passed overhead, the Avro rocked her wings in response. It was comforting to know that he was not completely alone.

Bo Beaudreau, an old trapper, and Herol, his grandson, stood transfixed as they gazed at the wonderous sight that passed overhead. The waning rays of light transformed the shabby plane into a translucent vision of beauty.

The Avro gentled into a shallow turn, and headed for the clearing west of the lake that was now fast receding into the evening shadows.

The ailing engine was nearly exhausted, it shuddered

once again, then belched a black plume of oily smoke into the solitude. The barking engine continued to protest as the Avro faded into the twilight haze.

Bo and Herol had strained their senses to the last moment attempting to see and hear the wonderous sight for as long as possible. Finally the last echo drifted across the lake shore. A thin blanket of mist hovered above the water, then silent nighfall. The old man and the boy trudged up the rocky path toward their cabin.

The meadow was faintly visible to Muggy as it emerged from the twilight. The tall grass on the field was a lush carpet from the air, but it might conceal rocks or stumps. Muggy didn't have time to make a pass. The engine was in agony, and the noisy fuel pump had all but torn itself apart but somehow kept feeding the thirsty Le Rhone engine.

He leaned against the rim of the cockpit, and as if by instinct, the weary plane banked its wings. The swish of air past the windshield and the singing wires hummed a comforting tune against the death rattle of the stricken fuel pump. The wheels skimmed over the pine tree tops, and a rapid ticking against the tires signalled contact with the tall grass on the field. Seconds later the shock cords clattered as the wheels rumbled across the rough ground.

Muggy's haggard features softened as the Avro skidded to an uncertain stop on the damp grass near the far end of the clearing. The engine shuddered once again and then died. He lifted his weary arms and rested them on the cockpit edge. His breathing was labored as he wiped the cold sweat from his grimy countenance. The acrid smell of burnt castor oil from the hot engine assaulted his nostrils and gave him a fleeting sense of nausea. He bent his head back in search of the pure night air and stared at the vast night galaxy overhead. Fiery diamonds, close enough to touch, sparkled their mysterious life into his soul.

Muggy was spellbound by the miracle that he'd been too blind to see before.

This ia an AVRO 504K. Muggy Pederson took Bo Boudreau and Herol on their first airplane ride in a plane like this. The plane is English and was manufactured early in World War I. (Cole Palen collection)

"Oh God! If Heaven is up there someplace, I'd sure like to see it. It must be wonderful, 'cause it's sure beautiful on this side."

His desperate eyes softened a bit when he realized that he'd said his first prayer. He didn't think he knew how. It was a good feeling he'd never known before.

In a split second the spell was broken. A harsh laugh emerged from the depths of his gut. It echoed the years of loneliness, fright and despair.

Finally, he lifted his bone chilled body out of the cockpit and surveyed his surroundings. A dense forest encircled the clearing like a black wall. He walked stiff-legged around the Avro, caressing the soft grass with his heavy boots. He changed direction abruptly and stumbled through the darkness toward the nearest trees in search of a campsite. Two large pines with overhanging branches grew nearby and would afford shelter for himself and the Avro. He trudged back to the plane and wearily pushed it under the protective branches. After the Avro was secured with tie-down ropes and safely bedded down, he dug into the rear cockpit and pulled a threadbare sleeping bag onto the ground. He unrolled it and spread it under the lower wing. An equally threadbare mosquito netting was tied to the inter-plane struts and draped over the bedding.

It was time for a well earned rest. He squatted on the damp ground beside the lower wing and fumbled in his shirt pocket for a sack of Bull Durham tobacco. His stiff fingers spilled more tobacco in his lap then on the folded cigarette paper that he held. He twirled the paper awkwardly with two fingers and licked the glued edge. A burning wooden match dispelled the inky night momentarily, then quickly died. He inhaled the pungent smoke, dispelling for the moment, the hours of anxiety he'd lived through once again. The night wilderness was his safe haven for the time being.

He searched through the pockets of his leather coat and pulled out a large chunk of dried beef jerky. He found a likely place to chew and gnawed hungrily at it. He stood up stiffly and reached into the rear cockpit for a bottle of whiskey. He moved over to his bed under the wing, stooped under the mosquito netting and sat down with the beef jerky and bottle between his boots. This was his hard earned wages for the day. Dinner at the Waldorf!

A comforting thought reminded him that his lifelong love affair with the Infinite Heavens would soon bring him home.

How empty life would be, he pondered, if this communion with God's universe had been denied him.

Top left—The author as a young boy.
Top right—His grandfather, Bo 'Gramps' Boudreau.
Bottom—His mother, Cecelia Salut Daley.
(Author's collection)

2 BIRDS OF A FEATHER

Old Bo and his grandson were still relishing the adventures of the day as they neared the end of the path that led to Bo's cabin.

Herol skipped a few feet ahead of the old man, chattering happily and reliving the magical sight still fresh in their memories.

"Gramps! That was an honest for real airy-plane wasn't it?"

"You bet it was! That was a real hum-dinger. That was one of them double-wingers," Bo replied.

"Gee! You're smart, Gramps! I bet you could fly one of them hum-dinger watcha-me-callits!"

"I reckon," Bo smiled.

Twilight had surrendered to the encroaching night. The boy sensed that Bo was weary and his knees would be getting stiff from the night air. He waited for Bo to catch up with him. He grasped the old man's hand in his chubby fist and they walked slowly the rest of the way to Bo's cabin.

"Let's sit a spell 'till I catch my breath, then we'll head for your house," Bo said.

They sat on the steps of Bo's cabin. Their senses were tuned to the abundance of nature's wild creatures that shared their wilderness realm. Fireflies played 'hide and seek' with their magic lanterns as Herol tried to catch them in his hand. Faint rustling in the brush near the cabin signaled the presence of small night creatures coming to life. A large owl glided past on silent wings enroute to its nest. Hundreds of tiny eyes glowed with reflected light from the moon as forest creatures played their role in nature's life scheme.

'Bo' Boudreau would be eighty-one years young with the coming of spring 'ice crackin'. He wasn't sure of the date, but his gentle Chippewa mother told him that on the day he was born the Great Spirit of Winter relinquished his realm. He commanded the Spirit of Spring to break the icy bonds that held the blue waters in deep sleep, and to breathe spring blossoms across the land.

Bo's father, a hardy French/Indian trapper was a practical man. He couldn't be bothered with legends of the Great Spirits. His life was survival in this wilderness land of serene grandeur, a land whose beauty masked hardship, loneliness, and frequently, sheer terror.

Beauregard 'Bo' Boudreau was a product of these simple, good folks. Today, he was Bo, the most respected woodsman, hunter, and trapper in the land.

Herol, his grandson, was a miniature copy of the old man. He lived about a hundred yards down the shore in a tiny cabin with his widowed mother, Cecelia. He divided his time equally between his mother and grandfather. They filled his days with tender love and legends of his ancestors.

Bo emerged from some reverie and said, "Best we get you home, your mother will have supper waiting."

"Gramps, you reckon that hum-dinger airy-plane will be around here tomorrow?" Herol asked.

"I reckon that flyer-feller was in big trouble when he passed over this evening," Bo replied.

"Gee Gramps! You're smarter than anybody! How can you tell?" Herol said with pride.

"Just figured he might be. I think he's bedded down in that clearing about a mile from here. I'll ask Cecelia if we can hike over there in the morning."

"Oh gee!" was all that Herol could say.

Bo clasped the boy's hand and rose slowly to his feet. They strolled leisurely down the wooded path toward the boy's home. A chorus of sleepy birds and scurrying night creatures announced their presence.

The old man knocked gently on Cecelia's front door before they entered. He kissed his daughter affectionately and Herol gave her a hug.

"It's even better than the picture books, Mother," Herol shouted.

"What are you talking about, son?" Cecelia asked.

"Honest for real airy-plane!"

"Gramps wants to take me to see it tomorrow. Can we go Mother — please?" Herol pleaded.

"Where son?" Cecelia looked up from her sewing.

"I reckon that feller set down about a mile from here, on the other side of the woods. There's a small clearing west of here," Bo interrupted.

"He flew over this evening and waved at us, mother. May I?"

Cecelia wished that she understood her son like Gramps did. Sometimes, she suspected that her father was getting soft in the head. And maybe it was contagious, because Herol was getting more like his grandfather every day. Those two could talk like a couple of fools for hours about things she never understood. Herol was a good boy and Gramps was her cherished guardian and provider since the death of her husband. So she tolerated the fantasies and outlandish talk they thrived on. Some day she would have to move to a larger community so that she could work to support herself and get proper schooling for Herol.

"Yes I suppose you can go," she said.

"Oh boy!" was the lad's reply.

"I'll be here by sunup," Bo replied as he reached for the heavy wooden latch beam across the door.

"Aren't you staying for supper?" Cecelia asked.

"No thanks, I'm pretty tired. Besides, I've had a kettle of bully beef and beans on the stove all day. I'm anxious to try it out. I'll bring some over in the morning."

Bo waved goodnight, unlatched the heavy oak door and stepped into the darkness of the night forest.

Bo was up before daybreak and on his way to Cecelia's cabin before the morning sun peered across the lake. A vigorous rap on the door announced his arrival.

"Come on in," Herol shouted.

"Good morning, Gramps," Cecelia smiled and handed him a tin cup brimming with hot coffee as he entered.

He acknowledged her greeting with a smile, and accepted the hot brew. In turn he handed her a black iron kettle.

"Them's my bully beef and beans, hope you like it," he grinned.

Cecelia thanked him. Then she gently chided the old man, reminding him that he should be teaching the boy the lore of the woods instead of foolishness like today's adventure. She reminded herself that a long talk was overdue. This would be the last time! She'd get it settled once and for all, right after this escapade with the fool and his flying machine.

Bo sipped the last of his coffee and the boy rose to accept the heavy-laden knapsack that Cecelia handed him.

"Coffee was good, I'll get Herol home by dark."

"There should be enough food in there for the both of you. There's extra in there for that fool flyer, in case you run into him. I expect he's hungry," Cecelia said.

"Thanks," Bo smiled.

She kissed the boy goodbye, and shooed them out of the door.

They walked through semi-darkness beneath the outstretched canopy of pine boughs. They trod silently on damp grass and pine needles. Occasionally, a faint ray of the morning sun lighted their path where the pine canopy parted. Soon the virgin growth of timber would give way to young seedlings and wild berries when they neared the area of second growth trees.

Bo led the way, and he'd kept a steady pace for a half hour. His stamina heavily taxed the boy's youthful vitality.

"How much further, Gramps?" Herol asked as his breath crystalized into frosty vapor.

"Not much," he grunted.

Bo stopped and waited for his young companion. "Best you let me carry that grub, it must be heavy."

Herol lowered his burden to the ground with a grateful sigh. A rotting log nearby afforded a comfortable seat for a short rest.

"If we find that flyer-feller, I'll bet he's hungry," Herol ventured.

"Mebbe so. Keep your eyes peeled for a spring, or maybe a patch of snow under rocks or broken trees. We'll make coffee later."

Bo picked up the napsack and swung it across his shoulder.

"Let's get goin," he said.

They trekked silently, peering expectantly into the white mist that enveloped them. A dense ground fog was forming as the early morning sun warmed the chilly air. They quickened their pace as the fog rose to form low hanging bits of cloud vapor.

Dimly visible toward the north end of the clearing, Bo saw what appeared to be a white awning. He increased his pace again as they approached closer. The vague shape took on the outline of an airplane tucked neatly between two large jackpines.

They spied a figure in a battered sleeping bag under the wings. He was blissfully unaware of the rain forest created by the wings above him as water vapor condensed and dripped in steady streams around him.

They gazed with awe at the wondrous machine in front of them. For minutes, they stood transfixed, their eyes never moved from the magnificent sight.

Finally, Bo set the napsack down carefully and motioned for Herol to open the pack and see what food could be shared with their sleeping comrade. There was coffee, a small pot, smoked venison, corn bread,and hard candy — ample to provide a fine meal for the three of them.

Herol gathered kindling wood while Bo stepped into the shadows in search of patches of snow to fill the coffee pot. He signaled the boy to build a fire several yards downwind from the plane near an outcropping of rock.

The coffee pot bubbled contentedly as it rested on a mound of burning chips. Bo poured a few drops in a tin cup to sample the brew. He was satisfied with it, so he filled the cup to the brim.

"Should we wake him up?" Herol asked.

"Dunno, maybe he'll smell the coffee. Get some cornbread and we'll walk over there."

They walked shyly toward the plane. Once again, the shabby plane was transformed into a magnificent illusion. The weatherbeaten fabric covering the wings and hiding numerous fractured wooden ribs, sparkled with crystal beads of dew. The wooden wing struts glistened with moisture, hiding their weathered condition. The propeller reflected the rising sun, transforming it to polished onyx. It was grand!

Bo stood in front of the sleeping figure, wondering what to do next. Perhaps the aroma of coffee stirred him, because he moved slightly, opened his eyes in bewilderment, then sat up.

"Morning, young feller! Thought you might like an eye opener." Bo offered the cup of steaming brew to Muggy.

"Thanks," he said. then there was an embarrased silence as they both grinned sheepishly.

"Herol give him some cornbread so he got something to go with his coffee," Bo said.

Muggy was ravenously hungry after his sparse supper last night. The unexpected treat was refreshing. He crawled out of his sleeping bag, stood up, and offered a hand to his unknown benefactor.

"Thank you. My name is Pederson, Magnus Pederson. My folks live on a farm about ten miles from here, just outside Bemidji. I was on my way there yesterday but I had some trouble with the engine."

"I'm Bo Boudreau, and this is my grandson, Herol. We live on the other side of these woods. We saw you fly over last night and thought you might have had trouble." Bo shuffled his feet, and didn't know what more to say, so he just grinned.

Muggy's reaction was akin to Bo. He offered him the tin cup. "Care to join me, Mr. Boudreau?"

"Sure! Call me Bo."

"Okay! I'm Muggy."

Herol stirred the fire, filled another tin cup with coffee, and ventured toward the men. This morning's events were beyond his wildest dreams. He stood silently beside Bo and offered him the steaming brew.

"Thanks son."

Bo took the boy's hand and drew him close, aware that he was shy. The great northwoods country afforded little opportunity for meeting outsiders.

The day waned all too soon. Muggy labored over the sick engine, repairing burned ignition wires, overhauling and cleaning the fuel system, and looking for all the reasons that might have given him so much worry yesterday. The fuel pump had been removed for a later inspection at Bo's cabin. Bo helped as best he could by staying out of Muggy's road, but alway available to reach for a tool or hold a part for him. Best of all, he lent needed moral support that Muggy welcomed. Herol spent hours transfixed in a wonderland he'd often dreamed of.

The afternoon sun rolled downhill on its descending arc toward the horizon. Soon the long northern twilight would usher in the evening stars.

Muggy reluctantly wiped the grime from his hands while he took one last look at his handiwork. Bo picked up the tools and stowed them in a box in the rear cockpit. Herol waited patiently, the sick fuel pump cradled in his arms. Soon they would begin the trek back home.

"I guess the old girl will hold together long enough to get home tomorrow," Muggy gave the Avro an affectionate pat.

Bo led the way toward home. Muggy carried the fuel pump and Herol carried the near empty napsack.

Twilight was minutes away when they reached the cabin. Herol took his reluctant leave, envious that Gramps would share the night with Muggy. Soon he was skipping down the trail that led to his home. He was excited to tell his mother about the wonderful events of the day.

Cecelia had a hearty meal waiting for the boy. Bo's bully beef and beans simmered on the back of the wood stove. An added treat, wild plum pie, fresh out of the oven, was on the table and still hot.

She greeted him with a gentle hug, then shooed him out to the back porch to wash his hands and face.

Herol was glad to be home with his mother. Her gentle ways always made him feel secure. He missed her after being gone all day.

He gulped his food down. The long day had tired him out. His mother sat nearby, she knew he was worn out with excitement. It would be bedtime soon.

The morning sun was still asleep when Herol pounded on Bo's front door. He was burdened with the napsack as before. Bo and Muggy were ready to leave, so without delay they stepped into the morning chill.

Bo took the lead as before and struck his usual brisk pace. Muggy took the napsack from Herol and handed him the lightweight fuel pump. They fell in line behind Bo. They walked in silence, each immersed in his own thoughts. The autumn morning chilled them to the bone.

The sleepy sun was skirting along the tops of the pine forest when they reached the plane. Muggy set the napsack on the ground and retrieved the fuel pump from Herol, while Bo searched for wrenches in the tool box.

As Muggy struggled with the repaired fuel pump, Bo and Herol built another fire and soon had the coffee pot whistling and steaming.

Repairs were completed in short time. Muggy blew his warm breath into cupped hands that were stiff with the cold. A warming cup of hot brew would be a welcome luxury in this morning chill.

Muggy was eager to get home. It couldn't be more than a ten minute flight from here. He hadn't seen his parents in over five years; they were lonely and anxious to see him. He'd written them about two weeks ago, so they were looking for him any day. Telephones were still scarce or he would have called. He upended his cup to savor the last of the hot coffee, then placed the cup near the campfire and stepped toward the plane. He was anxious to see if their efforts had proven successful.

He looked at the engine intently, looking, feeling, touching and looking again. He stepped on the lower wingwalk and boosted himself up for a final look at the fuel pump. He tapped the tiny propeller with his finger; it spun quietly and smoothly. He was satisfied that he'd done all that he could. When he got home, he'd dismantle the Avro in his father's barn and rebuild it like new over the winter months. For now it would have to do.

He stepped down from the wing and walked around the front of the craft. He swung the big propeller over several times before the engine belched flame and kept running. Muggy sprinted back to the wingwalk and reached into the front cockpit. He motioned for Bo to untie the rope fastened to the tailskid and climbed into the cockpit.

The engine hummed smoothly, but still had a subtle off-key ring that he had sensed yesterday. He let it run several minutes, and satisfied that yesterday's troubles were behind, motioned Bo and Herol to the rear cockpit.

Bo hoisted the boy into the cockpit. Then he stepped gingerly from the wingwalk to the cockpit edge, and slowly settled himself in.

Muggy gave the Avro full rein. The tail skid slid through the wet grass and held their course straight. As they accelerated, the rudder lifted off the ground, grass dragged against the wheels, and the propeller carved vapor trails through the damp air. Speed increased and eager wings lifted the frail craft and its adventurers into the autumn sky.

The roaring engine coupled with the tornadic blast of air past the cockpit kept the excited passengers breathless with a mixture of fright and bewilderment. Their fears were soon dispelled when Muggy turned his head and gave them a reassuring smile. They climbed slowly through wisps of morning fog until they reached the vast blue canopy above. The morning sun bounced off moisture laden wings and sparkled like diamonds. The checkerboard of greens, browns, and placid waters beneath them was a fairyland to behold.

The Avro leaned into a gentle turn that led them overhead the cabins. Cecelia was standing outside waving her apron. Bo and Herol waved and shouted their lungs out. These exciting minutes were equal to a lifetime! They had been inducted into the Realm of the Gods!

Muggy flew in a wide, lazy turn that eventually led them back to the makeshift airport. He began a long glide to the field, wires humming, and singing an ever lower key as speed dissipated. The wheels skimmed over the tall pines and they settled slowly toward the grass. A soft swish, the wheels ticked off tall blades of grass, a rumble, then the tail skid dragged a furrow in the ground.

They were earthbound mortals again!

Bo and Herol vied for Muggy's attention. They jabbered excitedly, neither making sense. Muggy just stood there patiently wearing a big smile. He waited for them to unwind.

Coffee was mentioned by someone during the verbal marathon, so Herol took the cue and scampered off.

Bo handed Muggy his cup, their farewell drink before his departure. It was time for Muggy to bid his new friends goodbye. He was grateful for their help and was reluctant to leave, yet home and family beckoned.

"Have to get goin' — anxious to see the folks. It's been a long time."

"Can't you stay a little while longer? I wish you could meet my mother," Herol said.

"See you again soon, won't we?" Bo put in.

"Me and the plane need a good rest. I thought I'd stay around until spring. Sure I'll be seeing you soon," Muggy smiled.

"Gee! That's swell. I wanna be just like you!" Herol blushed at his sudden outburst.

"One of these days I'll be too old to fly anymore. Then I'm gonna sit on the front porch and watch you fly."

Bo flung the remainder of his coffee on the ground and rose with an effort.

Muggy rose and patted the boy on the head. "Don't forget, I'm gonna watch you fly someday."

Muggy gave the propeller a hefty pull. A hollow bark and a plume of smoke brought the engine to life. He sprinted around the wing and clambered into the cockpit. Bo released his hold on the tail and Muggy sped across the grass.

"Isn't that the grandest sight ever, Gramps?"

"The grandest ever," Bo replied.

"Sacre bleu!" Bo muttered under his breath.

The plane had barely cleared the trees at the far end of the field when it faltered. The wings appeared to stagger as the nose dropped uncertainly.

"What's wrong?" Herol cried.

Bo didn't answer. A sense of fear gripped him as he watched the stricken craft. A dull thud, felt more than heard preceeded an oily plume of smoke.

"Muggy's in trouble!" Herol was frantic and clutched Bo's hand for reassurance.

They stood close together, watching silently, each alone in his thoughts. Doubt and hope vied for attention as they tried to share Muggy's peril.

The sick engine belched and gagged. The Avro seemed to hang motionless, its nose hanging at a precarious angle to avoid the trees.

"He'll make it," Bo muttered for his own benefit.

The plane staggered toward a cross-wind position and headed for a landing. They breathed easier, confident that Muggy would be able to reach the field.

"It can't be! Oh God!" Bo muttered and pressed Herol's face into his coarse mackinaw jacket.

A harsh bellow accompanied by a violent surge, caught the unstable craft in a deadly grip of engine torque. The wings mirrored the sun as the plane whipped into a sloppy roll and the nose dropped to near vertical.

"On no!" The boy tore away from Bo and ran in Muggy's direction.

Somehow, the gyrating wings stopped, Muggy was fighting vainly to bring the nose up when the airplane hit and sent a shudder through the ground as it exploded. The wings melted into a pile of burning rubble, the fuselage telescoped, then broke in two behind the cockpits. The nitrated fabric flared white hot and disappeared like celluloid, leaving the charred wooden frame exposed. A blackened figure moved sporadically, vainly trying to escape the flames. Then it slumped forward on its side exposing the charred skull of a once handsome man. Two cavernous holes stared into eternity while melted lips revealed a grotesque smile.

They raced toward the inferno, stumbling over each

other in their panic. Acrid fumes filled their lungs as they neared the wreckage. Waves of intense heat reached out greedy fingers for more. Their headlong scramble slowed to a hesitant step when the wind shifted, turning the fire aside revealing Muggy's skull grinning at them.

The boy paled to the color of chalk, and leaned on Bo for support. Bo stood rigid, his features set in disbelief.

They backed off from the fire as the wind shifted again. The stench of burning flesh enveloped them, and they coughed from the dense smoke. Herol covered his face with his hands and Bo led him away.

They sat or fell down in the grass, gulping the fresh air and coughing to rid their senses of the disaster. A dazed expression bespoke the agony in their souls.

"Why did it happen to Muggy? He was goin' home."

"Don't know son. I guess the Lord needed him."

"Couldn't He wait till Muggy got home?" Herol sobbed.

"Reckon not," Bo answered softly.

"The Lord didn't have to take him now. Muggy was goin' to sit on the porch and watch me fly someday."

Bo scolded the boy. "Don't you dare try outguessing the Lord! Muggy said he was going to sit on the porch — maybe he meant up there someplace. He'll expect you to fly someday a lot higher than he ever dreamed of."

"I ain't ever gonna fly again!" Herol countered angrily.

"Yes you will. You'll understand someday."

"Why did it happen, Gramps?" the boy sobbed.

"Maybe the good Lord sacrificed Muggy to make an example for younger flyers. There is much to be learned about flying, and Muggy's death might help a little."

"I don't understand you," Herol murmured.

"You will someday."

"Why did it happen?" the boy cried.

"Why?"

Herol (the French spelling of Harold) stands at the door of the hangar lean-to at Fargo, North Dakota. Inset — Front view of the Fargo hangar. Part of the lean-to can be seen on the right.
(Author's collection)

3 WHEELS 'N SKIS 'N STUFF

Bo had walked along the shore for most of the afternoon and listened to the awakening wildlife stirring from a long winter sleep.

He had spent most of his time either in his cabin or on his porch the last couple years since Cecelia and Herol had moved away. His favorite pastime was rocking in the ancient wooden rocking chair that Cecelia had given him before they'd left.

Days, months, and years had melted into one nameless segment of time. Life was empty and purposeless now that they had gone. Time was running out for him and he knew it. Accepting that, softened life's blows, eased many heartaches, and dulled the senses. Past, present and tomorrow all seemed to merge into one.

He nodded his weary head and decided, all in all, that life had been good to him.

The ice was fast receding from the shoreline and floating chunks of it collided in the mainstream and piled into a noisy mass. Snow drifts hidden in the forest's shadow tried to evade the warming breezes of spring, and the sun caressed the earth with warm fingers to coax out wildflower blossoms that would herald spring's arrival.

Was he eighty-two or was he ninety-two? It didn't

matter much. One year was as good as another. His grey eyes burned a little dimmer than years past but his silent reverie rekindled a spark when he remembered that Herol had promised to visit him this year come "ice crackin' ". He turned and headed back toward the cabin.

As he climbed the rock-strewn path, he smelled smoke from the cabin chimney. He didn't remember if he'd banked the fire or not. Maybe old Spider had come over to sit a spell. He opened the timber door and the aroma of hot coffee drifted through. His eyes adjusted to the dark room slowly.

"Hello?" he said.

"Hi Gramps! How 'bout some coffee?" Herol said happily.

Bo smiled. Herol had come home.

Bo's dim eyes lit up, "Was thinking of you!"

"You thought right. Come sit down."

He gave the youth a clumsy hug and sat down.

"Glad to see you. Going to stay awhile, I hope."

"I hitched a ride on a freight train out of Fargo, North Dakota, and the 'bulls' kicked me off at Grand Forks, so I thumbed a ride here. I can stay a couple days. Then I guess I'll be on my way to Fargo," Herol replied.

"What you doin' ridin' on freight trains?"

"I'm gonna try to get to Winnipeg, and then on to Saskatoon."

"Saskatoon?" Bo's face lit up with long ago memories.

"Might get a job there. I wanted to talk if over with you."

"What kind of a job?" Bo was alive now.

"Sort of a flying job," Herol faltered.

"Well don't that beat all! How about pushing that corn meal closer to the fire."

They sat down to the simple fare, savoring the hot food and taking comfort in their comradeship.

"How's your mother?" Bo asked, hoping to hear some news.

"Ok I guess, she married that real estate guy from Calgary, name's John Daley. She sent you a letter, didn't she?"

"Don't remember — is she well took care of?"

"Yeah, she's all right. I didn't know they got married until it was all over. I only seen them once since. We got a grass field in Fargo that might some day be an airport — I live out there in a lean-to."

"Sounds like you're on the right track." Bo tried to sound cheerful. "Now about this job in Saskatoon, or is it Winnipeg?"

"Everybody thinks I'm crazy when I mention airplanes."

"There's lots of crazies that think everybody else is crazy," Bo smiled.

"Mr. Daley thinks so, and she goes along with him too."

"Who's Mr. Daley — never mind, I remember."

"Guess you won't like this. I quit school over a year ago and been working at the airport. I couldn't get no sleep, so I quit school."

"What kind of nonsense is that!"

"I fell asleep in class a few times and they was going to expel me, so I quit."

"We'll get back to that later."

"I helped clear the trees to make the flying field, and built a shed for the planes. Dick Titus gave me a job when the first mail plane landed, and everything was jake until Mr. Daley tried to have us all arrested, so I figured to hell with Mr. Daley and school too — I been on my own ever since."

"I don't agree or disagree." Bo said kind of sternly.

"I got a bunk and a stove in the lean-to, and Dick pays ten bucks a month," said Herol.

"You can't finish school?"

"No. A mail flight from Grand Forks lands at midnight and I have to put out kerosene pots to make a runway and tend a bonfire so the pilot can find the field."

"Can't you sleep evenings or after midnight?"

"I tried, but it didn't work last winter — ever since I've had a sort of flying job."

"You flying?"

"Yeah — I have to put skis on the plane in Fargo, and along the line somewhere I might have to put wheels back on if the snow ain't deep enough."

"Where you get all these skis and wheels?"

"We got pairs of them stashed away along the route and extra skis in case we bust one. It's a pretty good plane too."

Bo wanted to ask if it resembled the plane that Muggy was killed in, but instead he asked, "Is it like the ones in picture books?"

"It's something like the one that Muggy got killed in."

Bo seemed relieved that time had softened the hurt of Muggy's death.

"Does the pilot let you do any flying?"

"Not much chance, I have to sit on the mail sacks, and it's mighty cold behind the propeller — guess I'd be too frozen to fly."

"Where you eat and sleep?"

"I don't except at Pembina. There's a shed with hay in it. I don't eat much but I take a sack of grub along. I try to bum coffee when I can."

Bo was immersed in his own thoughts. He knew the boy was seeking approval, but this was no ordinary job, it was a fantasy of the future—the same dream that claimed Muggy's life. No wonder Mr. Daley raised a fuss.

"How many of these skis and wheel things you done?"

"Maybe a hundred." Herol felt proud of his accomplishment.

"You might get all busted up on one of those flights or maybe get lost in a blizzard." Bo was trying to fill in the blanks that were missing.

"It's pretty hairy, and I'm tired of being cold and

hungry, but I'd regret it the rest of my life if I quit."

"You figure to sit on them mail sacks and fool with them skis till you get killed?"

"I don't know. I gotta get experience."

"If you live long enough. Ain't there any other jobs in this flying business?"

"A few guys do stunts at the county fairs — and they're as hungry as I am. Some are trying to start airlines, but the railroads are too much competition."

"I got a sort of flying job."

"Thought that's what you was telling about."

"Yeah, but this is another one. It's something like Muggy did, except it's run by a businessman, and he hires these pilots and other guys."

"You're not getting a job as a pilot, so you must be one of them other guys."

"I met a guy in Winnipeg one day when we was weathered in. We were in the cafe down the road from the airfield — we got talking and he recognized me as one of the guys on the mail plane. He was going back to Saskatoon to make arrangements for the spring tour of the flying circus he belongs to. He needs a partner to replace the one that got killed. He promised lots of real flying and maybe next year I'd get on as a regular pilot." Herol's brain was as muddled as the story he tried to tell.

"Is he one of them crazy guys that does tricks?"

"That's what he does — he's famous too — they call him Diablo on the advertising posters."

"How come his partner got killed?"

"He jumped out of a plane in a parachute."

"You going to jump out and get killed too?"

"You said this was a better job than Muggy had. I think it's a danged sight worse! At least Muggy got killed *in* his airplane. Guess you already made up your mind." Bo mumbled to himself.

"Guess so, I hoped you would understand."

"I'm trying to, but you're leaving too many blanks in

your story. I'm going to tell you a story, then you make up your mind." Bo's eyes mirrored a wistfulness that Herol had never seen before. His fierce glint submerged into a soft forlorn expression.

When I was about fifteen years old I had dreams too. I wanted more than hunting and trapping. The sea beckoned me just like the sky calls for you. I never saw a great sailing ship except in pictures. Them sails remind me now of the wings on your airplane. The sea offered me something the woods never could."

"I thought you loved the woods, Gramps," Herol interrupted.

"I told my parents, and the more I talked the less they understood. I could no more explain those white sails than you can the wings on a plane.

"Why didn't you tell me all this before?" Herol asked.

"Wasn't nobody's business 'till now. I set out on foot for Mystic, Connecticut. That's where the big whaling fleets and clipper ships sailed into for repairs. I was hired aboard the Sea Witch, a beautiful three masted cutter, as a sailmaker's apprentice. I wrote home every week and told of the fine life I had. The letters from home told how lonely they were, and how the trap lines were being neglected. After thinking about it, I decided to return home and help my parents. It took me four months to get back, and by then I'd made up my mind to be the best woodsman in the country — it was all I had left to hold onto. My father died within five years, and my mother followed him a few years later. By then it was too late to start over, and I've regretted it all my life."

Bo talked more this afternoon than in years past. The sense of dignity that he inspired would be forever etched in the youth's mind.

"Thanks — you've said it all for me. I might get killed, but I gotta try."

"I might have got lost at sea. Now listen to me! The Heavens belong to the Almighty, just as the sea is part of

Herol in downtown Bemidji during one of his visits with Bo 'Gramps' Boudreau.
(Author's collection)

His realm. He might welcome you, or He might resent your intrusion. It's up to you to decide if the rewards you hope to gain are worth the price you may have to pay." Bo leaned back in his rocker and closed his eyes. In a minute he was asleep.

It seemed to Herol like a lifetime ago since he'd last seen Gramps. So much had happened in the last ten years. Herol was doubtful if it had all turned out for the better.

Their lives had been turned topsy-turvey. Nothing would ever be the same again. The quiet solitude, the unhurried tasks of the day, the love and security of family, were all gone.

Cecelia was living in Fargo too, along with her new husband John Daley. They eeked out a precarious existence by toiling at any menial tasks available. Life was hours of work for the two of them. Their reward was a life shared with each other.

Herol's survival was not a great accomplishment. His job at the airport provided him with the barest living. Making do, getting by, was a problem that gnawed at him each day. Survival on the ground meant filling his gut with a pot of rice—it was cheap—or a pot of beans. Beans were more filling but also more expensive. Survival in the air was an unknown quantity. Ambition, fear, uncertainty was all he could offer.

How many times had he been tempted to turn his back on his dream and return to the peaceful serenity of the north woods? Gramps had done it for his parents. Why couldn't he?

Times were tough all over though. This was the great depression and there were few jobs for anyone. Soup kitchens and bread lines were a part of life. Men desperate for work turned to a hobo existance in search of work. They rode the freight trains in whatever direction they happened to be traveling. There had to be a job somewhere.

Herol looked at Bo sleeping in his chair. He could see he was growing feeble. His body had grown thin and his once-sure step now faltered. The rugged woodsman of years past had vanished with the passing of time. His family had moved away and now his only companions were the forest creatures that he knew and loved. There was his longtime neighbor too, old 'Spider'. Spider was another ancient creature of the forest.

What to do? Herol recalled Gramps story.

"The sea beckoned me just like the sky calls for you," Gramps had said. Herol knew then what he would do. He'd follow the call just as Bo had wanted to do but couldn't. There was no one to call Herol back. He would follow his dreams of the sky. Gramps would want it that way.

They spent three more days together. Great days walking the shore of the lake watching the northland come back to life after a long winter. They talked some more, mostly about the past, but some about the future. Herol never did tell Gramps what he decided. He didn't need to. Gramps knew.

Then on the fourth day, after sad and embarassed farewells, Herol left for Fargo. Gramps watched him on the road until he disappeared.

A Douglas M-2 Mailplane. Herol rode along and helped out when Kiddie Karr flew the mail in an M-2, from Fargo to Grand Forks, Pembina and Winnipeg, Manitoba, Canada. Herol's job was to change wheels for skis and vice versa whenever needed.
(Author's collection)

4 KIDDIE KARR

A faint shaft of light glowed from the side window of the wooden hangar that sat forlornly on the Fargo airport. It cast a dim yellow glow on the mud and half melted snowdrifts. A shadow passed by the window, and moments later, the tin stovepipe on the roof of the lean-to belched a shower of sparks.

Inside, Herol paced across the room, swinging his arms and blowing warm breath into cupped hands. There was a chill tonight, and the exercise helped a little until the fire came back to life. The stove's warm glow felt comforting, and by taking measured steps across the floor, he soon revived his chilled body.

A swaybacked bed and a two-burner kerosene stove stared at him from the bedroom wall. The stove's shaky legs straddled a wooden box that served as cupboard, pantry, and catchall. A battered wooden chair leaned against the adjacent wall, and beside it another wooden crate supported a decrepit alarm clock that clanged away the hours. A small window over the bed faced the landing field, now a soggy field of mud and melting snow.

His weary eyes stared at the barren room without seeing it; he picked up the letter from Bob Tyson postmarked Minneapolis, Minnesota, and read it once more.

Hi Kid

I don't write many letters — I'm settin' here in Minneapolis having a big party. I been here a month and almost broke.

I'm going to Saskatoon in two weeks, will stop in Fargo on the way, if you still want the job we talked about. I'll come on the Great Northern Railroad if I'm not broke — otherwise, I'll ride the rods on the Northern Pacific. And if I miss you, will get in touch somehow.

Happy landings,
Bob Tyson

Tyson must have forgotten to mail the letter for a few days, Herol thought. It's only a week before he shows up, if he shows up! His head nodded as sleep overtook him, and the letter dropped from his hand.

The ancient clock groaned, then clanged as the hands jumped to 2:30. The noise shattered Herol's sleep. He groped for the jangling alarm and the wily old clock fell to the floor and bounced off to a remote corner of the room still rattling its tin bell. The never changing routine added one more scar to the battered timepiece.

He rose with a start as his senses tried to shut out the offensive noise. He discovered that he'd been sleeping in his clothes. So much the better, since he didn't have to dress, that allowed extra time for a sip of coffee.

The morning sun was still fast asleep as he shuffled into the workshop, kicked open the iron door, and stirred the dull embers in the stove. He gathered a handful of kindling wood and threw it in, stirred the fire again, then added a split oak log.

The coffee pot lid was open and he added a dipper of water to yesterday's grounds. He filled a five gallon oil can from a barrel against the wall, then placed it on top the stove beside the coffee.

He pulled on a sheepskin lined flying helmet and picked up a pair of leather mittens and stepped through the doorway into the hangar. "Kiddie" Karr would fly the M-2 mail plane to Minneapolis at daybreak and more than likely wouldn't return in the near future.

He checked over the canvas engine cover that hung to the floor, pushed a gasoline heater into the tent-like enclosure and lit it.

He would miss Kiddie Karr. They had shared many hectic flights together over the past winter months. They'd become comrades of necessity and, later, friends by choice. He hated to see him leave and could only hope they would cross paths again. He'd show Tyson's letter to him and see what he thought about it.

He recalled the first time he met Kiddie. It was several months ago on a frigid night in early winter. Gusty winds and blowing snow foretold that blizzard conditions would be a certainty when Kiddie was expected to land. It was to be the first night flight from Minneapolis inaugurating the expansion of air mail routes further north and westward out of Fargo.

In spite of the storm, the city fathers, government officials and businessmen crowded into the hangar lean-to as they awaited the arrival of Kiddie Karr.

A telegram from Minneapolis had been their only communication so far. The expected landing time, 9:00 p.m. was given and as an afterthought, the author also volunteered the information that the weather was awful!

Eyes and ears strained to catch the first glimpse or sound of the Curtiss mail plane as the clock in the lean-to clanged nine bells. Minutes later, the laboring drone of an airplane engine was heard, only to fade out as the wind carried the sound in another direction. Visibility was zero, and swirling snow devoured the darkness and hid any star that might attempt to see through.

The snowbound field was shrouded in darkness except for a bonfire made of engine oil and wood, and a row of kerosene pots. Hopefully, the bonfire would be Kiddie's beacon to guide him to the field. The kerosene pots strung in a line just beyond the fire marked his landing direction. And how he would ever find the hangar in this storm if he got down at all, was a minor problem to be faced later.

After what seemed an eternity, the sudden swoosh of fabric wings was heard as they flashed across the bonfire and hovered above the kerosene pots, then disappeared into the void beyond. Minutes later, an engine bellowed above the roar of the gale winds as Kiddie struggled to keep the big biplane on an even keel. The whirling propeller emerged through a cyclone of snow as the plane staggered in the general direction of the hangar.

Herol remembered how important he'd felt when he'd stepped in front of the onlookers to signal the pilot to the hangar. Then he drew his mittened hand across his throat indicating 'cut engine'. He greeted Kiddie, helped him unbuckle his parachute harness and assisted him out of the cockpit.

They leaned against the harsh wind and labored toward the warm shelter of the hangar. Close to their heels was the horde of welcomers. They all headed for the door inside the hangar that led to the hot stove and coffee pot in the lean-to.

Kiddie Karr was enormous in his bulky leather flying suit, fur trimmed goggles and helmet, boots, and gauntlet mittens. And midst a crowd of jostling wellwishers, he elbowed enough space around him so that he could shed his warm flight clothing.

He tugged off his helmet and goggles revealing thin straw colored hair, pale blue eyes set far apart, and a face divided by a long sliver of a nose. His thin lips were devoid of color and lifeless. A long pointed chin completed his rather bland appearance.

Kiddie's skin would have been the envy of any woman — creamy white! Not a blemish showed across his plain features except for a multiple array of tiny furrows that converged at the corners of his eyes. Numerous chalk white lines zigzagged in haphazard fashion across his cheeks and jaw.

His head appeared too small for his huge body until he stepped out of the bulky flying suit, revealing a thin, loose-jointed frame, and misshapen fingers.

Herol's big disappointment came when Kiddie fumbled awkwardly into his shirt pocket and produced a pair of rimless eyeglasses that he carefully adjusted on his nose. And, Herol's thought at the time was that "he looks like a confounded school teacher!"

That had been months ago and now Kiddie was leaving.

After one last gulp of coffee, he turned off the gasoline engine heater and retrieved the hot oil from the stove in the lean-to. He poured the oil into the big M-2s oil filler, capped it and hopped down.

Grunting and cussing, he managed to get the hanger doors open far enough and finally he manuvered the big biplane out the door into the cold morning air.

Now for the fun part, he thought. He chocked the big biplane's wheels, set the mag switch and throttle in the cockpit and ran back to the propeller. The hot oil did the trick and on the third pull, the big V-12 started.

Just as Herol climbed into the cockpit to do a little ramp-flying, a Model A rounded the corner of the hanger and pulled up next to the plane —it was Johnnie Sims, the mail truck driver, and Kiddie Karr.

"Good morning!" they shouted in unison. Johnnie jumped from the front seat, grabbed a mail sack, and walked around the wing and threw it in the front compartment.

"Hot coffee on the stove," Herol shouted over the engine.

"No thanks, gotta get back," Johnnie shouted back.

"I'll have a cup," Kiddie called.

Herol nodded and climbed out of the cockpit. They left the engine of the M-2 idling as they walked to the hanger.

Inside, Herol poured them both a cup of the brown water. He drank his in silence.

"Something on your mind?" Kiddie asked.

"Yeah, here!" Herol said handing him Tyson's letter.

"Do you know Tyson?" Kiddie asked reading the note.

"Not very well, I met him once last winter — ain't never seen a guy like him before."

"I know what you mean — he's an odd-ball character. I've known him a long time. Can you keep a secret?"

"Guess so."

"Me and Tyson were flying buddies in the Hollywood Black Cats — ever hear of them?"

"Sure — a hot flying circus," Herol was proud of his knowledge of flying lore.

"I quit because I got badly burned in a crash, that's why I'm such a weird looking guy now. The medics built me a different face, so I started over again — mail pilot this time."

"Looks like I'm gonna do the same thing — maybe it ain't a good idea," Herol said.

"What you go to lose?" Kiddie asked.

Herol paused, "I might get killed."

"I damned near killed us a few times," Kiddie smiled.

"That's different — guess I'm scared 'cause that guy Tyson gives me the creeps."

"Life isn't much fun if we don't gamble a bit. You got any other offers?"

"Hell no! I ain't got nothin' nobody wants."

Kiddie glanced at his watch. "We'd better get goin'."

He stepped out the door. As they approached the plane, Herol asked, "What should I do?"

Kiddie put his foot in the stirrup and hoisted himself over the cockpit edge, then turned around.

"I live to fly — I don't fly to live! What's your reason?"

"Me too!" Herol shouted as he waved farewell.

His wistful gaze followed the uncertain path of the M-2 mail plane as Kiddie jockeyed the biplane through a sea of mud and melting snow drifts. The laboring propeller spattered the wings and open cockpit with a shower of mud and flung a dark, sticky spray behind in its wake.

Finally, Kiddie reached the downwind end of the field and stopped. He wiped the mud from his goggles with the sleeves of his leather flying suit. His teeth were gritty, so he spit over the cockpit edge several times. He made a last minute inspection of his parachute. He checked the three emergency night flares beside his seat. Another last check of his safety belt, and a glance at the engine instruments, and he was ready to go.

The engine roared, another geyser of mud spray engulfed him. He ruddered the plane into the wind, as he flipped the ignition switch from 'left' to 'right' magneto, then 'both' for a final check.

The M-2 rolled slowly ahead, wallowing through the muddy tracks in front of it. Kiddie held the tail low to the ground to diminish the spray picked up by the propeller. Speed increased ever so slow but steadily. As if by sheer courage, the M-2 hoisted itself out of the mud and slithered across the surface for half the length of the field.

Slowly, under Kiddie's expert guidance, the big mailplane lifted from the mud and water. Kiddie was still low and gathering speed as he came over the hangar.

He looked out over the side of the cockpit and waved farewell to Herol. Herol partly waved and partly saluted in return. Then he stepped to the corner of the hangar and watched the biplane climb off toward the horizon where dark roll clouds were forming ominously.

Herol walked back and through the hangar door and into the lean-to recalling the many blizzards he and Kiddie had flown through since their first meeting. They could have been killed any number of times. And now Kiddie was heading off into another blizzard.

Kiddie wasn't a movie hero pilot, he was a real hero—a gentle, quiet, unassuming man. He was a giant among his fellow pilots. He was truly a gentle giant.

A photo of 'Doc' Bestine's Travelair biplane. Dick Titus and Herol flew this airplane frequently. In the right corner, part of Herol's Henderson motorcycle can be seen. (Author's collection)

5 LIVE FOR TOMORROW

Dick Titus made a hurried trip from his downtown office to the Fargo flying field. The month's inventory of gasoline was due, and he was confident the kid would have the figures available. He made a mental note to argue the city fathers out of a few more dollars for Herol. Those penny pinchers thought ten dollars a month was plenty. He braked his car to a halt near the hangar and saw the youth kneeling beside the gas pit measuring fuel. "I'll damned well get a five dollar raise for him if I have to pay it myself!"

Dick was in a vile mood this morning, just like every other morning. Living was a big bunch of crap if he had to exist like this. He'd been offered this job a year ago while he languished in the hospital nursing two mangled feet. The bills were piling up, and he couldn't join up with his flying buddies. Besides, his trusty Travelair was busted up worse than him. He'd do "a little flying and a little paperwork" — that's what he was told when he took this job. To date, his flying was near zero and the paperwork was piled to the ceiling.

He was glad the field didn't have a telephone, it gave him an excuse to drive out more often. It was good to smell the familiar scents of a hangar again — engine oil, fresh nitrated fabric, acetone, the tar and cinder ramp. The planes smelled especially good! That office would drive him nuts after a taste of the peaceful silence and clean smell of the grass field.

He had reasons for the dark moods that clouded his nature. His feet still bothered him, and one fracture had knitted imperfectly and left him a semi-cripple. His bouncy walk of yesteryear was replaced with an awkward shuffle that reminded him he was a "has-been".

Before he broke himself and the Travelair into pieces, he bounced from one day to the next on short stocky legs and laughed at the world with twinkling blue eyes. Years of hardship hadn't left a mark on him until now.

"Hi Dick!," Herol greeted.

"Hi! How's it goin'? I gotta sign for the gas we used last month."

"We're short again this month, maybe I forgot to mark a few gallons down."

Dick scanned the totals rapidly and with an impatient gesture — "To hell with it! Let's call it shrinkage and make it up next month."

"Shrinkage" was the late evening raids to fill Dick's Hudson, and Herol's slate wasn't clean either — a few gallons in the rickety Standard whenever he had a chance to do some early morning flying. He knew what to expect next, so he might as well ask.

"You want some gas in Doc's Travelair?"

"Good idea! By the way, you still thinkin' of that job with Bob Tyson?"

"Yeah, he's supposed to be here in about a week. Guess I'll try it. You know him?"

"Everybody knows that clown! He's wilder than hell! He'll make you or break you! I guarantee that!" Dick's face lit up with happy memories.

"He's headin' for Saskatoon after he stops here."

"Maybe I can get a ride for you guys as far as Winnipeg. There's a new Hamilton comin' through in a few days."

"Swell!"

"I gotta get back to town, can't fly today — fill 'er up just the same." He turned on his crooked foot and self-consciously eased himself into the car. The happy expression of moments ago was replaced by drawn features and desperate eyes.

A week later Herol awoke with a feeling of apprehension. All heck could break loose today. Tyson might show up by plane, train, or whatever — if he didn't forget! No time to pack any junk, no goodbyes, and Dick might not show up either.

The Great Northern scheduled a train to Fargo daily around noon, and the last word from Minneapolis was that "Cash" Chambers would fly the new Hamilton in around two o'clock. There should be enough time to meet the Great Northern, then get to the freight yards in case Tyson rode the rods, and still time to meet the Hamilton.

He pushed the old Henderson motorcycle out of the hangar and was about to fire up when he heard the familiar chug of Dick's Hudson rounding the corner of the hangar.

Dick waved as he drove past, "Help me crank the Travelair."

"Confound it! I'll be late for the train," Herol mumbled.

He leaned the Henderson against the wall and ran to the grass area behind the hangar. Dick was standing beside his car, arms laden with packages.

"This stuff might tide you over until you can scrape a few dollars together."

"Thanks!"

"Open the stuff," Dick prompted.

"Honest to God new socks — and no holes in 'em!"

"Open the rest!"

The next bundle contained a pair of heavy shoes and leather mittens. Dick reached inside his coat pocket and handed him an envelope.

"Payday — or almost. Four extra bills in there for a vacation, just in case things don't work out with Tyson. You might decide to come back. Take them seedy shoes off — the socks too — how many pieces of cardboard you got in them shoes?"

"Three or four, I guess," Herol was embarrassed standing with his toes and heels exposed.

"Get them socks off!" Dick grabbed the worn shoes, spun them over his head, and let go with a reasonable semblance of a pro ballplayer.

"The shoes squeak, that's supposed to be good, ain't it?"

"Sure kid, now let's get the show on the road." Dick was smiling as they walked toward the plane.

"Crank this turkey. I'm gonna wring it inside out today! I'll be back in time to meet the Hamilton."

He was a different man. He stood taller, and the awkward limp seemed to have righted itself. He bounced into the rear cockpit, pulled on his helmet and goggles, and hollered "Contact!"

"Contact!" The OX5 engine sputtered to life.

He gunned the engine, and the lumbering biplane gathered momentum across the grass. Dick waved a happy goodby and was gone.

The Great Northern's whistle screamed as it signaled its approach to Fargo. He could make it to the station in about ten minutes if the Henderson didn't let him down. The gravel road was rutted with piles of loose shale, and the rear wheel whipped from one rut to another as the tire spun off the rocks. He crouched low to avoid the debris, intent on keeping the bike upright, when he saw a flashing red light.

"Damn! A ticket." He slowed to a stop, kicked out the parking stand, and sat astraddle the bike.

The black Dodge touring car pulled in front of the

motorbike and 'Big Tim' Reilly eased his bulk out from behind the wheel.

"Hi Kid! You got a black and yellow airplane at the field?"

"Yeah, the Travelair that belongs to 'Doc' Bestine."

"Well, Doc's in trouble. A farmer about five miles north is hollerin' that some crazy guy is tearing up and down his pasture chasin' his cows. They're scattered all over Hell's half acre."

"Doc hasn't been around for a few weeks. I think he's out of town."

"Anybody else fly it?"

"Me and Dick Titus do once in a while."

Herol was certain the officer wanted him to go back to the field, and rather than get trapped, he volunteered — "I'll get in touch with Dick on my way to town; he'll look into it right away."

"Well, uh, that might be all right since you're headed that way — but I'm holdin' you responsible. Where you goin'?"

"Gotta meet the Great Northern, guess I'm late now."

"You help me and I'll help you," said the policeman. "Follow me and I'll get you there."

The Dodge spun its wheels in the loose gravel and kicked up a cloud of dust in its wake. Herol gunned the Henderson to a roar and hung on for dear life as he groped his way through the dust.

Tyson stepped gingerly off the train. His head was splitting; he needed a shave, and his blue serge suit should have been fumigated. He approached the small sandwich counter inside the depot, gazed wistfully at the pretty girl in the Coca Cola advertisement, then at the wilted waitress busy swatting flies.

"What you got to drink?"

"Strawberry pop."

"Nuts!" He turned and walked out the door.

If the kid ever received that letter, he should be here. If not, he'd board the train for Winnipeg. He had a lot of fuzzy weeks behind him, and it was an effort to remember what to do next.

A police car and motorcycle bore down on him as he walked alongside the depot. "What kind of happy crap is this? I ain't done nothin' in this town yet," he growled.

Big Tim braked the Dodge to a halt a few feet in front of Tyson and Herol pulled alongside.

"Thanks, he's my passenger."

"Don't forget, have Dick get in touch with me right away." The patrolman waved and backed away.

"Sorry I'm late, got trouble at the field." Herol extended his hand in greeting.

"Hiya kid, glad to see yuh."

Herol took Tyson's battered suitcase and tied it to the rear fender of the bike. "Have a good trip?" he asked.

"Slept most of the way, feel pretty rough."

"We got a chance to deadhead to Winnipeg, if you think you can make it. Hop on!"

Tyson straddled the bike and off they went.

As they approached the field, they saw several cars parked at random around the small wooden structure that served as mail room. Herol glanced in the direction of the hangar, hoping that he'd see the Travelair.

As they crawled off the motorcycle, Big Tim strode toward them. His features were grim.

"What gives with these cops?" Tyson growled.

"Shut up, don't say nuthin' — I'm in trouble."

"Where is Dick Titus? You been coverin' for him?" asked Big Tim. "I got a mind to run all you nuts in jail. Why didn't you tell me that plane was missing? Is *he* flyin', or who is?" The barrage of questions waited to be answered.

Herol stammered a few unitelligible words, only to be

cut off with another blast.

"I got a bunch of reports about this nut. Some say he hit a tree, another that he's still raisin' hell, and another that he's dead! I got a mind to run you in. You stay close by, understand?"

"What in hell is goin' on?" Tyson growled.

"I think Dick busted up the Travelair."

"Hey! There go two cop cars, let's follow 'em." Tyson pointed toward the road.

They trailed the police cars until they approached a circle of curious spectators. Several cars and wagons were parked alongside the road and in the field. They bounced across a ditch and rode through a stubble field toward a stand of cottonwood. As they dismounted from the bike, the air was rent with a series of angonized bellows, followed by the staccato bark of two pistol shots. They elbowed their way through the crowd in time to see a patrolman replace his smoking gun in its holster. A wild eyed heifer lay half buried under the engine of the Travelair, still jerking spasmodically in its death throes. The plane teetered at a crazy angle, the top of the rudder dangled above the ground, and the collapsed wings were driven toward the rear of the fuselage. The impact of plane and beast impaled the heifer with the broken tip of a propeller blade and disemboweled it.

Dick hung half way out of the cockpit, his arms disjointed, and buried in the rubble. One twisted leg was jammed hard against the windshield, and his neck swiveled back at an extreme angle burying his face in the ground. His safety belt was unbuckled, by intent or accident. Only Dick knew.

Herol's eyes glazed as he stared at the misshapen figure in the cockpit. He diverted his eyes, only to be confronted with the sightless stare of the dead heifer. He felt giddy, the ground spun under his feet, he sank to his knees and vomited. Harsh visions of Muggy's grinning skull in the burning Avro flashed back — and now there was Dick!

"Forget him! The sooner the better! Don't think — ever!" Tyson's voice was hard.

"You're crazy! He was my friend!" Herol shouted in anger.

"He was a friend of mine too. Now he's dead! Understand?"

"No, I don't!"

"You will! To hell with yesterday, it's gone. And to hell with today, it's nearly over. Live for tomorrow! Now let's get out of here, we got a plane to catch."

"You damned spook! You ain't human." Herol muttered as Tyson grabbed his arm and walked him toward the motorcycle.

They rode in silence to the field. The Hamilton, a big metal monoplane rested in front of the hangar. They pulled alongside the sleek craft and parked the bike in the shade of its wing.

They stared at the corrugated metal wing, their eyes following its graceful span that seemed to extend to infinity. The big Hornet engine shimmered as hot vapors rose from the cooling cylinders, and the exhaust manifold crackled as the metal contracted. A large greenhouse of a cockpit emerged from the leading edge of the thick silver wing and extended almost to the engine. The helmeted and goggled pilot would be out of place inside this weatherproof cockpit. Passengers could sit in comfortable wicker chairs at their individual windows. It was a truly magnificent creation.

For a brief period, the Hamilton helped to dispel their gloom, then stark reality returned when Officer Reilly strode toward them. He was tired, and the aftermath of the day's tragedy showed on his face. His usual hardboiled expression was gone.

"Hey kid — don't forget, I might need to get in touch with you in case anything more comes up about Dick. You'll be around here someplace, won't you?"

"Yeah sure."

"Nice looking plane — be seeing you." Big Tim departed.

"Let's get goin' and find out about hitching that ride," Tyson muttered.

"Shouldn't we postpone everything — maybe Dick's funeral?"

Tyson gave his companion a baleful look, and after a long silence, "None of that buryin' crap for me — I'm gonna remember him when we flew all day, chased broads and boozed all night. And that's just for today, tomorrow I'll forget him!"

Herol was startled and was about to open his mouth when he was interrupted.

"Hell, I ain't even goin' to my own funeral, hope there ain't enough pieces left to bury. Listen kid, forget em' fast or all you'll have to think about are bags full of busted bones. Now let's find out about that ride."

They climbed the wooden steps leading into the mail room. Cash Chambers would likely be in the cubby hole off the main room — the radio shack. They saw his huge frame bending over the radio operator's shoulder as they listened to the static-laden message coming through from St. Paul. He straightened up as he accepted the handwritten message and stepped into the outer room.

"Gee! You're sure duded up! You look swell!" Herol extended his hand in response to the huge paw offered in greeting.

"Glad to see you, Herol."

"Hey Cash, I'd like you to meet Bob Tyson. We're tryin' to bum a ride north."

"Glad to meet you, Bob — heard a lot about you, glad we've finally crossed paths."

"Heard about you too." Tyson felt ill at ease, aware of his dog-eared appearance compared to the elegant blue and gold uniform that Cash wore so handsomely.

"Gee, you fly like that, no helmet and goggles?" Herol asked in wonderment.

"Why sure! We'll all fly like this in a few years — a real modern pilot for a modern plane, that's me!"

"Where's Dick?" Cash asked.

A pall of gloom enshrouded them as they searched for words that wouldn't come.

"Hoped we could freeload a ride with you," Tyson snorted through his handkerchief.

"Why sure! We seldom have any passengers. Dick was supposed to be here when I landed over half hour ago."

"He creamed himself," Tyson said flatly.

Herol fought back tears and stumbled past them and through the door. The ghosts of Muggy and Dick jeered him as his instincts led him to the only safe haven he knew, the decrepit lean-to. An hour or more must have elapsed when he awoke. He was summoned back to reality by a rough tugging on his new shoes.

"Get up. We're gonna be late!" Tyson hollered.

He rubbed his swollen eyes, wiggled his toes in the new shoes, then sat up. He remembered Dick's last words — "Let's get the show on the road!" That's damned well what he'd do! Tyson told him, "Don't think!" He'd better do that too.

"Okay, right with you. Help me pick up some of this junk."

"Hell! It's all junk!"

"Yeah, I know — just anything that looks worth takin'. I'm gonna grab the coffee pot and stove and a couple pots. Guess we'll need them wherever we're goin'."

They struggled through the hangar with their odd assortment of ragged clothes, foul smelling stove, pots and pans. By contrast, the shimmering Hamilton stood aloof waiting its motley passengers. They settled themselves into the two forward wicker chairs behind the cockpit. Cash gunned the big Hornet engine and they taxied out on the cinder ramp.

Herol looked out the window at the hangar and lean-to as they taxied past. So long Home Sweet Home he thought.

No telling what's in store up ahead.

The big Hamilton rumbled and bounced around as Cash turned it into the wind. Then he slowly pushed the throttle in and the plane began to trundle forward. It was deafening in the cabin and Herol put his hands over his ears. The tail came off the ground and the takeoff run smoothed out slightly. Then the wheels lifted, touched back down, and then lifted for good.

For better or worse they were on their way.

Herol pressed his nose against the window as he watched the receding ground fade away. The clean sky held no terror for him. It helped him forget the harsh world below. It was a puzzle to him. How could the world be so beautiful all the time from up here and be so hard and miserable much of the time on the ground.

Tyson might be right. To hell with yesterday! It's gone. To hell with today! It's nearly over. Just live for tomorrow!

A pilot and official inspect the skis on a big Hamilton Mailplane. Herol and Tyson flew from Fargo to Winnipeg on a Hamilton flown by Cash Chambers.

6 CANADA BOUND

Tyson soon lost interest watching the fertile Red River valley glide beneath them. The Hamilton's sparse interior did little to muffle the roar of the big engine as the vibrations danced along the corrugated metal skin and converged in his aching head. The past several weeks of "booze and broads" was about to catch up with him, and the noise magnified the demons that pounded in his head. He leaned back, stretched his legs another inch and closed his eyes. He'd deprived himself of sleep too long.

Thoughts of Dick flashed across his mind but he quickly dispelled them. It wasn't a good idea to have friends; they had a habit of getting killed. The hurt was always there, and when it gnawed too hard, the sure fire remedy was a trip to the bootlegger and total oblivion for a few bleary hours.

It must be fifteen years since he'd drawn a completely sober breath. He was either drunk or sobering up. The fuzzy interlude between was reserved for his best aerial performances. It was no secret to him that he was a misfit, admired for his uncanny luck, but socially unacceptable.

"Well nuts !" he thought. "Time is running out for me just like it did for Dick today." Most of the old gang were gone. A few made spectacular exits like Dick. The unlucky ones vegetated in wheel chairs.

He opened his eyes to see Herol leaning against the back of Cash's seat. They both talked at once. Cash pointed and Herol's eyes followed. The kid was probably driving him nuts with questions, but Cash seemed to be responding with enthusiasm.

As he watched, the seed of an idea germinated in his brain, an idea that if planned properly, would give the kid a break and satisfy the conflict that raged within his own being. This idea would truly be a tempter of Fate for both of them. The kid would go on to better things than being an airshow bum. He'd make him into a respectable pilot, like Cash for instance. He decided to lay off the booze at the right times, and keep out of the hootchie kootchie joints, and look after him proper.

Cash Chambers hummed a nameless tune as he winged alongside the placid Red River. This was a far cry from the "hell for leather" days flying mail across the Rockies. He looked over his shoulder to see how his passengers fared. Tyson was asleep, and Herol was scribbling on a postcard. He wondered what would become of this oddball pair after he left them in Winnipeg. A chill ran through his big frame at the thought of Tyson setting an example for the youth. He decided that a pep talk was in order before they parted company.

He pushed back his natty gold braid cap, rolled up his sleeves, and bellowed out another song.

"Herol! Ain't this the life? Beats hell out of working!"

"Who wants to work!" Herol shouted back.

Tyson stirred from soggy dreams, shifted his legs, and recommenced snoring.

Herol scribbled the last words on the postcard. It was a quick note to Gramps with no mention of Dick's tragic end. A long letter would follow when he arrived in Saskatoon.

"Hey Bob!" Herol shouted to his sleeping companion.

"Yeah." Tyson answered, half awake.

"Big Tim said I should keep in touch with him.

Suppose he'll send out a warrant or something for me?"

"Who?"

"The cop back in Fargo."

"Phooey on him!" Tyson said and attempted to go back to sleep.

"He'll throw me in the pokey!"

"Don't worry, I'll take care of you. Before the summer is over, we'll be run out of more towns with the sheriff on our heels than you can count." He yawned, and settled back to a peaceful snore.

On they flew toward Winnipeg somewhere over the horizon.

The Canadian sunset held out a last ray of red-yellow light across the Winnipeg sky. Street lamps sparkled in the early dusk, darkened windows mirrored a red glow, then turned silver as interior lights of the homes and office buildings came to life.

The bus depot was a nondescript hulk of greystone on the exterior and equally uninviting inside. They faced a huge blackboard on the wall that announced the bus schedules. Departure and arrival times were scrawled in yellow chalk, written only for the benefit of the unknown author.

The Saskatoon bus was due to depart in minutes. Cash would return to his beautiful Hamilton and go on to greater triumphs, the new partners would embark on an aerial odessey that would bring excitement, anguish, and more than they bargained for.

They sat in the booth killing time, empty cups in front of them, unspoken thoughts between them. Cash turned to the boy and suggested he go check the departure time. So Herol wandered off.

"Listen Bob. Give this kid all the breaks you can. He's a good kid and deserves more than he's had so far. He's in fast company for his age, so don't ruin him and make an air show bum out of him. Make a pilot out of him, and get him out of that stuntman stuff as soon as you can. You've seen too much and done too much, flying and otherwise. The otherwise is what I'm thinking of."

"I agree. He ain't in good hands, but I'll do the best for him that I can. I'm about at the end of my rope, maybe he can do a better job where I leave off." He grinned his clownish best for Cash's benefit.

Cash looked at Tyson. The comic expression unnerved him, it was the mask of a jester. His eyes mirrored a tortured soul awaiting release.

"A couple minutes before we leave," Herol said coming back to the booth. He felt little enthusiasm for the trip ahead.

Midst the din of pots, stove, pans, and bundles, they struggled toward the waiting bus. Cash hoisted the heaviest bundle over his shoulder and walked between them.

"Hey Cash! I forgot to mail my postcard — ain't got no stamp, but got some pennies here someplace." Herol was busy talking to himself.

"Never mind, I'll mail it for you. You guys have a good trip."

"We're gonna see the end of the rainbow," Tyson said as his lifeless eyes drilled into Cash. It was an eerie feeling, but somehow reassuring. Maybe the kid would be all right.

Suddenly the "All aboard!" They were gone, the farewells too hasty, thoughts left unspoken, and the dull ache of friends departing.

The driver scowled as they rattled and clanked through the doorway. They could distribute their conglomerate of junk under the rear seat, and judging from the haughty stares of their fellow passengers, hide themselves as well.

The night sky had overtaken evening twilight, and the air was heavy and oppressive, a sure promise of a rainswept journey. The driver clashed gears as he wheeled the silent busload of passengers from the depot into the night traffic. Street lights reflected their progress on the wet street as the driver jostled his way between jaywalking pedestrians and nervous drivers.

Tyson assumed his customary position, his legs stretched under the seat in front of him, and head slumped on his chest.

Herol was alone with his thoughts. The day had been too long and too full of memories. He thought of the money in his pocket and his new shoes — all a part of today — and Dick's happy face when he bounced into the cockpit of the Travelair. The hurt in his eyes was hidden by darkness as he looked at Tyson and wondered what made this guy tick. He'd been told, "Don't think, live for tomorrow." — If Tyson lives only for tomorrow, there's never a today for him. That's like living a fantasy! Maybe he ain't real!"

Herol folded his flying suit and laid it across the stove to form an uncomfortable pillow, the pyramid of pots and pans were held in loose array between his knees. An orchestra of sound floated through the aisle with each movement of the bus. Echoes of sleepy protest completed the noisy ensemble.

Daylight was horizons away and the moon was still holding to the night as they neared the outskirts of Saskatoon. The driver summoned his restless cargo to attention with a lusty cry of "Saskatoon", accompanied with a flash of bright overhead lights throughout the bus. Weary passengers squinted in the unaccustomed light, and slowly made ready for their exit.

"Watch me get a free meal and a few hours of shuteye when we get there," Tyson whispered under his breath.

"We ain't broke, I got over fifty bucks!"

"Shut up! Wanna get somebody to pick your pocket?"

"I got a few bucks too, but we gotta stretch it until we hit the road. It might be five or six weeks before we see any more green stuff."

"What we gonna do?" Herol asked.

"Stick close to me, and don't say a word!"

The bus pulled into the rain drenched depot to discharge its passengers. They waited till the others departed then stumbled through the aisle with their noisy possessions. The rain pelted them and soaked quickly through their clothing as they labored across the wet pavement.

"There's a couple of live ones! See the old guy and the dame at the lunch counter? If they're eating breakfast, we'll set beside them."

"Yeah sure — but can't we sit in a booth instead?"

"Damn! — not a word!"

"Where we gonna put all our junk?"

"Drop it on the floor!" Tyson dropped his bundles, then pulled Herol's arms down releasing the tin clatter on the floor for all to hear. It was Tyson's fanfare as he walked on stage!

"Let's go!"

Tyson's shoulders drooped noticeably, giving his rainsoaked suit an even more threadbare appearance. Herol followed close behind as he was told. Tyson assumed a solemn expression as they straddled the stools beside the elderly couple.

A sleepy waitress dropped a menu in front of them and turned her back to pick up two glasses of water.

"Whatta ya have?" the sleepy one asked.

"One cup of coffee, ma'm — for both of us," Tyson said weakly.

"OK, two cups of coffee."

"No ma'm, you don't understand, one cup of coffee, for both of us."

"One cup for both of you?" The sleepy one woke up.

"Yes ma'm," he said weakly.

Herol felt a blush spreading across his face. He hoped the coins in his pocket wouldn't jingle.

"OK! You guys get one cup of coffee, and you don't get no refill either 'cause I ain't goin' to give you two cups for the price of one!"

"That's all we can afford." His voice was complete dejection as he glanced toward the old couple beside them.

The kind old lady held her fork in midair, hot syrup dripping from the dainty morsel of pancake. She nudged her husband and whispered, then cast a furtive glance over her shoulder.

"Here's your one cup of coffee, and pay me now."

Tyson handed her the coin, "You take the first sip, son."

Herol blushed cherry red, but managed to do as he was told, tasting the coffee, then setting the cup on the counter. Tyson reached for it and took a gulp, then handed it back. He glanced wistfully at the kind old lady and then studied the bus schedule on the wall.

"Excuse me, Sir. Are you and your son traveling far?"

"We'd like to get to Calgary, ma'm," Tyson replied politely.

"My husband and I couldn't help overhearing your conversation with the waitress. We'd be so happy to help you in any way we can. Please be our guests for a hearty breakfast."

Tyson let it be known that he would be forever grateful. Herol continued to blush and said nothing.

"You hoped to go to Calgary? Perhaps we might help," the kindly benefactor beamed.

"Well it's like this . . ."

"It's settled then! You order your breakfast and we'll buy your tickets to Calgary. They hurried over to the ticket agent on their errand of mercy.

The waitress took their order for "two tall stacks with bacon", and when she placed the hot dishes in front of them, she scowled, and then laughed aloud!

They interrupted their breakfast to accept the tickets from their benefactors, and with a profusion of thanks, Tyson waved an affectionate farewell as they departed.

"What'd I tell you?"

"You told me to keep my mouth shut!"

"I told you we'd get free breakfast and some shuteye."

"Yeah, but how come we're goin' to Calgary?"

"Hell! We ain't goin' to Calgary!" Tyson snorted.

Herol was more confused than ever.

"I figured that expensive luggage by the door belonged to them, the baggage tickets said Saskatoon. They musta come in on the bus ahead of us."

"But where are we goin'?" Herol insisted.

"We ain't goin' nowhere! Them tickets give us a place to sleep till morning!"

"I ain't sleepin' on no bus!" Herol replied.

"Me neither kid! While we wait for the Calgary bus, we sleep right here in the depot. They can't kick us out if we got tickets!"

"That's crazy!" Herol shouted.

"Not so loud!" Tyson cautioned.

"Why did you let those folks buy us tickets if we don't take the bus?" Herol felt dumber than ever.

"Damn! We cash the tickets in come morning, and we got enough folding money to buy eats and a hotel room for a week!"

"It sure ain't honest. Wish I was back in Fargo — always had a place to flop, and could usually bum something to eat when I had to."

Tyson's expression turned grim. "Cash made me promise to take care of you. Well, by damn! I'm showing you how to survive when the going gets tough. What we did wasn't honest, but if we were flat broke it would seem honest as hell then. There's a good chance one of us might get killed in a plane, but the chances are a lot better that we'll starve to death first."

"We ain't . . ."

"Shut up! We might be dead broke lots of times, and we won't be pretendin' then."

"I thought we were goin' to do lots of flying and make a heap of money."

"We'll make a heap if we're lucky, but all we gotta do

is bust one of us up, or the plane — that gets expensive! In the meantime we gotta eat."

Herol reflected how simple and direct it used to be when he talked to Gramps, and what a puzzle this Tyson was. He not only talked riddles, he lived them, because he was one! The days ahead seemed to offer nothing but uncertainty.

"Remember what Cash said?"

"Guess so."

"About flyin' — remember?"

"Sure! — This is the life, beats hell out of workin'!"

"Still believe it?" Tyson asked in a kindly tone.

"I'd sooner fly than eat. If we miss a couple meals, we'll fly instead!"

"You got it kid!" His tortured soul receded into the depths of his lifeless eyes. A momentary flicker of peace shone through as he looked at the smiling youth's face.

Top: Herol 'Kid Galahad' Salut; 'Pop' Geraud; Carter Camp.
Center: 'Pancho'; Russ Sorensen.
Bottom: Eddie Gardner; 'Digger' McGonnigle and 'Gimpy' Gaines; and Harry Ross.
(Author's collection)

7 THE EAGLES GATHER

As dawn pushed a sleepy sun above the horizon, dark rain clouds swallowed the faint glimmer of light. The Saskatoon morning promised to be a soggy affair as the night's chilly drizzle persisted a while longer.

Harry Bottomlee scowled at himself in the bathroom mirror as he carefully drew a straight razor across his pink jowls. His glance alternated from the mirror to the living room windows of his apartment. The low hanging clouds obscured the view of his bank across the street, and threatened to deprive him of the meager light that filtered through.

Harry Bottomlee was very meticulous, whether shaving, dressing, or counting his money. Everything to the smallest detail must be absolutely correct. This morning he was up at dawn as usual, and annoyed because the soggy weather didn't fit into his orderly routine.

He was considered a shrewd businessman who could smell an opportunity long before his competitors got the scent. His interests encompassed the financial spectrum of banking, real estate, oil, lumber, cattle, and in recent years, aviation.

His success may have been in part attributed to the impression he created—an easy going fat little man, with an equally fat bank roll for the taking. Although his hand-tailored suits, and well-groomed appearance contributed to an aura of success, they still couldn't hide his angelic countenance. His pink bald head sported two small blue eyes, and chubby red cheeks surrounded a small mouth and equally small voice. His thick neck somehow managed to squeeze into a white starched collar, and his expensive suit draped over excess poundage to conceal a pair of stubby legs. He was neither liked nor disliked; he had no close friends, few social ties, a man sufficient unto himself. His shrewd mind was aware of his shortcomings, and like other opportunities, he profited by them.

Airplanes, and the men who flew them held a peculiar fascination for him. His business sense told him there was money to be made if he could tame the strange breed of uninhibited characters he'd grown to know.

In 1925 he'd taken his first ride in a wire and canvas contraption that was nursed into the air by sheer luck and a gutsy pilot. Another world was revealed to him in that brief interval, and a newfound admiration for those who lived in it. It never occured to him that he should learn to fly, he simply wasn't the type. Yet, his business sense and eagerness to associate with this new world aroused his curious mind. After that first ride and seven years later, much satisfaction had been realized from his venture, although little profit.

He finished shaving, patted his pink skin, and looked once more in the mirror. Tomorrow he expected his boys to show up for their first get together of the season. It was time to get organized for the flying circus that would tour the provinces and the states along the U.S./Canadian border. They probably wouldn't all arrive today in this weather, but he was confident they would all show up within a day or two. Some would fly, others would drive their broken down jalopies, a few might bum their way on the freight trains.

He could picture the flyers sauntering in. They'd stand,

sit, squat, or lay down — whatever mood prevailed. They were unruly because of their independent life, and always flamboyant to hide the strain they lived with. Most would be penniless, and they'd eventually get around to wheedling a cash advance from him. These aerial saddle tramps would never amount to anything, but they had one redeeming factor. They were paving the way for a future generation of flying respectability.

The expected chaos tomorrow morning made him tense and unsure of himself. Instinctively, his left hand reached for his bald head, and nimble fingers massaged it to a blushing pink. He rose from his desk still massaging his head, his right hand reached down and vigorously scratched his fat bottom. The realization of his comic posture made him blush and he quickly clasped his hands together. A recurring thought, year after year, returned this morning. He was always referred to as "HB". It had a certain mark of respect, yet these aerial vagabonds also called him "HB", and their use of his initials had taken on vague overtones a long time ago. He hadn't resolved this mystery yet. Perhaps some day he would.

He had to stop that confounded scratching!

"Now listen! Do like you did in the bus depot—don't say a word. I'll do all the talkin'. Almost forgot, you're eighteen years old in case ol' Happy Bottom asks you. Don't you volunteer nuthin'," said Tyson.

"Happy Bottom?" asked Herol.

Tyson clamped a hand over Herol's mouth, "For cripes sakes. Don't ever call him that anywhere he might hear you. It's just a little name we have for HB. You'll see why in time."

Bob Tyson stepped quickly to the closed door of HB's office.

"Hey! HB! Anybody home?" He gave the door a sharp rap.

"Me and my partner is here — I wrote you about him."

Hurried footsteps crossed the carpeted floor. HB's benign countenance failed to reveal the shrewd eyes that examined the two early arrivals.

"Good morning! Come in."

"Mornin' HB, this is Herol. Uh, Herol, this is Mr. Bottomlee, our manager."

"Glad to see you again." HB extended his hand in greeting to both of them. "A pleasure to meet you Herol." He gestured toward the conference room as they stepped through the door.

Tyson held a firm grip on Herol's arm and gave him a sly wink as they sat down.

"I'm surprised to see you gentlemen so early in the morning, considering the foul weather. Would a cup of coffee be tempting? My secretary will be here any minute and will bring some to us, I'm sure."

Tyson eyed the cherubic HB and reminded himself that flyers were dollar signs in this wily old banker's eyes. He'd have to work fast to get a decent break.

"We got here a day early so that we could talk a separate deal with you before the rest of the gang showed up. We got a whole new act worked up. Still needs to be smoothed out a bit, but it's along well enough to give you a preview if you got a plane available. Have you got a rebuilt Standard or Dehavilland that we could lease for the season?"

"I'll get in touch with 'Pop' Geraud. He's taking an Eaglerock out of winter overhaul today."

"Thanks! How about a practice session? And at the same time we'll give you a sneak preview what the suckers got in store for them. We'll make the yokels swallow their tobacco and the girls pee in their drawers!"

Herol lounged comfortably in the leather chair until Tyson's startling remark jolted him to attention. He gulped, blushed, and hoped nobody noticed.

HB eyed the strange pair sitting in front of him. They were as unlike as a cat and mouse. Did the wily Tyson intend to use the boy as a come-on, or was he really on the level? Time, or fate, would provide the answer.

"I'm prepared to offer the usual percentage of the gate receipts and a guaranteed minimum same as last season. As for the lease on an airplane, maintenance and repairs are your responsibility."

Tyson's icewater eyes glinted across the table. "We got here early so we could make a private deal. We're good enough so that we can sell to the highest bidder, and we're givin' you first chance — so a separate deal, or no deal at all! We guarantee this new routine will be a killer!" Tyson grinned at his own morbid humor.

A chill shivered down HB's neck. He felt a strange uneasiness as he looked at this living caricature who bargained with him.

"Very well, but not before I've seen a sample of this new show — and it will have to be good, otherwise no separate deal." With that, HB turned his attention to Herol who'd listened to every word with bug-eyed amazement.

"What do you think of all this Herol? Do you want to negotiate a separate deal also, or are you in agreement so far? And by the way, how old are you?"

The unexpected barrage of questions brought forth little more than an unintelligible stammer. Tyson took the initiative with, "Yeah, we talked it all over; we go 50-50 and he's eighteen years old."

"Yes sir, we're 50-50 — and I'm eighteen," Herol repeated.

"Thank you, Herol," HB said. "I was afraid your partner would monopolize the conversation. You look young for your age, I can't afford to get in trouble, you know. I'd say you were fifteen, at the most." HB smiled his angelic best.

"Cut out the crap! We both say he's eighteen!" Tyson growled.

HB knew better than to press the issue further. "All right, agreed, he's eighteen — and we put it in writing!"

The door opened quietly as an attractive secretary carried in a tray laden with coffee, cups, and saucers.

"Didn't I tell you we'd have coffee in a few minutes!" HB smiled proudly at the girl and introduced her. "Miss Savageau, my secretary."

She dutifully acknowledged the introductions with a nod of her head and a captivating smile, then placed the cups on the table, filled them, and departed.

"Boy! That's one classy lady!" Tyson grinned.

Stoney silence from HB and another blush across Herol's face was the only response. It was Tyson's turn not to press the issue.

The partners gulped their coffee, mindful of the free breakfast they had in the bus depot. HB savored the aroma and sipped delicately from his cup, aware that he'd better eat a hearty breakfast tomorrow morning before the remainder of these characters showed up.

"Suppose we get together again tomorrow. You might let Miss Savageau know where you can be reached, and she'll take care of the details."

They were on their feet before he'd finished speaking. A last gulp of coffee and a quick handshake acknowledged their commitment to an impossible challenge ahead.

Early the next morning HB hurried along the wet pavement as fast as his chubby legs would carry him. It would be good to feel the warmth of his office again. The thick slab of bacon and fried eggs sat comfortably in his fat stomach, and now he had the strength to deal with the other vagabonds who would soon confront him. He hoped that Miss Savageau would be able to keep them in line. No doubt they had driven her to distraction by now.

The front entrance of the bank revealed a spacious marble and glass lounge, crystal chandeliers cascaded from the lofty ceiling to cast a soft light on the oil paintings and statues that adorned the walls. He was justly proud of his treasures. These signs of success represented many years, and now he relished showing them to his clientele.

The usual number of early morning customers were transacting business. A few waited in line, while others browsed, admiring the art collection. Three unfamiliar figures caught HB's eye as they huddled in a far corner enjoying their own company far greater than his art treasures. As he came closer he recognized the trademark, leather boots, breeches,and leather jackets — a trio of aviators on hand for this morning's session. He was pleased to see such enthusiasm but just as quickly became annoyed when one of them used his lovely painting to demonstrate some insane flying adventure. The animated tale was enhanced with gestures that used the scenery in one of the pictures to zoom over time and again. He decided to ignore these ruffians and stepped into his private elevator.

Miss Savageau greeted her boss with a grim expression. She pointed to the closed door. "They're in there, about a dozen. Shall I announce your arrival?"

The low murmur of voices drifted through the closed door, punctuated with peels of laughter. HB prayed that they wouldn't have their boots propped on his mahogany table.

"No thanks, I'll just walk in," and with a sigh of regret and a sympathetic look from the girl, he opened the door to a tumult of greetings. He wore a fixed smile, nodded to the right and left at familiar faces as well as a newcomer or two. He raised his arms in a gesture of silence as he strode to his chair at the end of the conference table. He looked down its length at the collection of faces. Some faces were young, some old. Some enthusiastic, some cynical. Some returned year after year, others disappeared into that flyer's limbo only God knows where, and some were killed flying.

He went through this formality every spring. It was a

chance for the men to size each other up and his opportunity to look them over too. By the end of the session, they would no longer be strangers but would've taken the first step toward becoming a tightly knit group.

Miss Savageau opened the door slowly and pushed a cart laden with pitchers of coffee and cups into the room. Midst low whistles and murmurs of approval, she placed the coffee and cups around the table. A few minutes to relax, then HB would call the shots.

Downstairs, the three flyers HB had passed were lost.

"Hey Pancho! Where is the conference room?"

"I dunno. What's a conference room?"

Pancho grinned at his buddy, 'Dutch' Hauptman. "This flying is gettin' crazy, si?"

"Where in hell did Eddie go?" Dutch growled.

"I dunno!" Pancho shrugged.

Eddie Gardner was the brains of the trio, or rather, he was depended on to get them out of their numerous scrapes. True to form, he found the conference room and was hurrying back to retrieve his bewildered buddies.

"Let's go, fellas!"

They bounded up the wide staircase to the mezzanine, ran down the corridor, and were promptly brought to a halt as they stumbled into Miss Savageau's office. She politely ignored them as all three began talking at once, and handed them white cards to fill out. She pointed to chairs lined against the wall.

HB's coffee break was about to end. He'd enjoyed this brief respite from business by chatting with his old friend, Carter Camp, who sat next to him. Carter was one of his year-round employees, and had worked for him ever since that first airplane ride — a million miles ago.

With a friendly gesture, HB indicated it was time to begin the introductions around the table.

First there was Don "Gimpy" Gaines, a stuntman who

originally hailed from Mobile, Alabama. How he ever managed to end up so far north was a well-kept secret that only "Gimpy" knew. As he rose to face his companions, he grasped the edge of the table and leaned heavily forward. A loud creaking, followed by a muffled thump, brought a crooked smile over his weathered features. He rapped his knuckles against his right thigh, revealing to all present that a wooden leg substituted for the original that ended somewhere above his knee.

Ten years ago when he was flying the night mail across the Rockies, his tired old Liberty engine threw a rod and let him and the DeHavilland down on a mountain top. Two years in a hospital and one year feeling sorry for himself was the result. If he couldn't use the wooden leg as a pilot again, he'd fly anyway. He was a stuntman. Stuntmen didn't last long, but they crowded weeks into days — he'd been at it seven years! It was all borrowed time!

Miss Savageau opened the door when the murmur of voices subsided. Before she could utter a word, the latecomers that she'd held at bay scrambled past her. Several of their buddies were awarded pats on the back until HB's frigid look caught their attention. Miss Savageau directed them to three vacant chairs and without a backward glance withdrew to her office.

HB offered a flat good morning to the trio, then directed his attention to the brash young man seated next to Gimpy.

"Digger" McGonnigle bounced to his feet and gave the entire room a toothy smile. He was a huge blonde bull, muscular to the extent of being awkward, and fitted with a voice that rasped like a buzzsaw. "Digger" was probably about twenty years old, though he affected the mannerisms of a much older man. He claimed to be a pilot, and must have been the oldest pilot on record considering the dozens of exploits he laid claim to.

His Christian name was James, although "Digger" had

been tied to him when it was discovered that his first job was as an assistant to an undertaker near Klamath Falls, Oregon.

HB gave a friendly nod to the weathered looking occupant in the chair next to Digger.

Johnnie Day's (he preferred Captain Day) claim to fame was his four exciting years spent in the Royal Canadian Air Service during the great World War. His star rose briefly during this period when he led his squadron of Sopwith Camels on daily flights over enemy lines in France. His adjustment to civilian life was a bitter pill, and he was never able to settle down. The heroics of yesteryear were relived over and over again when he performed his solo extravaganza of that bygone time, complete with rockets, fireworks, and aerobatics.

He owned a tired old Hisso Standard that he tinkered with before and after every flight. The bright varnished struts barely hid the dry rot in the wood or their worn sockets. Numerous coats of paint added an ounce or two of strength to the rotted fabric, and suspicious bulges were probably old fractures that healed with the benefit of tape. With all its blemishes, the weary Standard looked for all the world like an aging burlesque queen, no longer young, but still beautiful with all that paint.

During the winter months he disappeared. Rumors had him as far south as Latin America. There were always uprisings, revolutions, and new generals taking over — a brief taste of glory for Captain Day!

Without prompting, Pancho, who sat next to Johnnie, rose to the occasion and introduced himself.

Pancho was a mystery to everyone, including himself. All that was known about him was that he started life as a street urchin in Tijuana, Mexico. The name Pancho was bestowed on him long ago for want of a real name by an itinerant barnstormer who befriended him and took him along as mascot. Barnstorming was better than begging in the streets, so Pancho grew up in the only family he ever knew.

"Dutch" Hauptman rose to his feet while Pancho was still grinning and shaking hands all around. He was Pancho's sidekick and had taught him the mysteries of the airplane and how to care for it. Dutch was getting old, probably near seventy. He'd had his share of flying in younger days, but as he grew older his inquisitive nature led him to tinkering with planes more than flying them. He'd become an artist in fine tuning the cantankerous Curtiss and Hisso engines.

Aside from his love for engines, a second love affair, and equally demanding, was booze and women. Whenever he stepped a few yards away from a plane, one or both would carry him to oblivion for days until a posse scoured the gin mills and hootchie-kootchie joints to rescue him.

Eddie Gardner, third member of the trio, turned as he rose to face HB. He offered an apology for being late — they were admiring the sights downstairs and didn't realize it was so late.

HB managed a brief smile and thought to himself, "Using my lovely paintings to demonstrate their damned crazy flying!"

Eddie Gardner was probably the only barnstormer that carried two mechanics with him. It wasn't always that way. He considered himself a fair mechanic until one day "Dutch" happened along and found the gremlins in his engine that had him grounded for a week. Eddie and Dutch had been partners ever since. One day, Dutch took off on a wild safari that lasted for several days and Pancho covered for him. That's when Pancho was considered a permanent member of the crew. It was a good life for the three of them.

"Black Jack" Reeves was the biggest, loudest, and meanest looking guy to ever sit at this table. His huge bulk gave the impression that he was sitting alone, although the table was crowded on all sides of him. There were those who'd been the target of his wrath and would swear he was meaner than a pit full of rattlesnakes.

Somewhere, somehow, James Milligan Reeves acquired the name that fitted his personality — "Black Jack". It was a reasonable guess that in the dim past he claimed a French-speaking province as home, judging from the faint inflection of his words and phrases that he used during unguarded moments. His name surely wasn't common in the provincial vocabularly, but names change just as people do.

His weary Standard was tied down in the pasture next to Johnnie Day's old relic. It too was a reminder of better days—another garish paint scheme that tried somewhat unsuccessfully to hide the ravages of time. "Black Jack" was emblazoned on the scarlet fuselage in bright yellow letters — upside down — telling the world that his aerial magic was performed inverted.

HB enjoyed these sessions immensely although he never gave the slightest clue. These rough-shod characters were ill at ease amidst the elegant decorum of his conference room, and the impromptu introductions, speech-making, if you could call it that, put them under a strain. These valuable minutes gave HB a chance to size up each man, and his first impressions were seldom wrong. He gave a friendly nod to the weary looking newcomer seated next to Black Jack.

Bernie Tannenbaum was a mousy little man recently arrived from Brooklyn, New York, via freight train, hitch-hiking, and walking. He arrived in Saskatoon three days ago, and slept or shivered through the cold nights in an empty boxcar. His "traveling money" had long since been spent, and he was down to his last five dollars of "eating money". He'd cut it pretty close, but that's the way he planned it.

He wore thick-lensed glasses that magnified his eyes and gave the illusion that he was continuously surprised at everything and everybody.

His former job as a presser in a garment factory netted him a fat paycheck each week, and Saturday would find him on his way to Roosevelt Field, Long Island, New York. Although he had money for flying lessons, he never got around to making that first flight. When he inquired about lessons, he wasn't seen — or maybe they ignored him, he wasn't sure. It was like this all his life — people weren't aware that he existed. He felt like a bug in the woodwork.

Bernie was here today through sheer guts and several letters he'd written to HB attesting to his prowess as a parachute jumper. Hopefully, HB wouldn't try to verify names, places, and events too soon, or maybe never, if he got started on the road soon enough.

Today he was in Saskatoon midst this flying fraternity—and best of all, he was one of them!

Bernie Tannenbaum — parachute jumper!

HB studied this strange little character and made a mental note. He'd talk to him in private when he had the time. Something didn't ring true. He dismissed the thought for the time being as another of his boys stood up.

Harry Ross was an ex-carnival barker, con artist, and, in recent years, front man for HB's flying circus. He was the perfect man for such a role — brash, smooth talking, and a Dapper Dan with the ladies. He was completely devoid of conscience. Promises were made to be broken, and responsibilities were a nuisance. He could think fastest when under pressure — and that was most of the time. Frequently, he'd be carried away with his own ballyhoo when his imagination went on a rampage. He'd promise the city fathers of every community an air show extravaganza — and then worry how he was going to produce. The worries usually ended up in the laps of the barnstormers, and they'd break their necks trying to live up to his promises.

On too many occasions the local sheriff would be after him, and when he couldn't be found, the planes were impounded, or worse yet, any available body became an unwilling guest in the town jail.

Every flyer knew Ross, or eventually would. For all his

faults he was a wizard! He made money flow into everybody's pockets, including his own.

He wasn't a pilot, nor did he show the slightest inclination to becoming one. Harry considered himself a super-salesman and a respectable thief. To use a quote of his — "I'd steal my grandma's eyeballs and sell 'em for grapes!"

That was Harry Ross!

HB stood up, paused a moment for silence — "Gentlemen, two of our members not present this morning checked in early and were called away on pressing business. I didn't want any of you to think that Bob Tyson wouldn't be here this season. Some of you recall the unfortunate loss of his partner last summer. He has a new partner this year, and you'll probably see Tyson at the field tomorrow with his promising new stuntman."

He nodded then to the handsome blonde giant sitting next to Harry Ross.

Russ Sorenson was a big Swede from southern Minnesota. He was on his way up in this flying game — lots of hard work, a love of flying, and a few lucky breaks put him far ahead of many of his older companions. The circus was a summer vacation for Russ. He'd usually stay most of the summer and then return to his regular job, flying the mail from Los Angeles to points east. He represented what so many aspired to but would never realize. Everybody hoped that Russ would make it all the way.

Rene "Pop" Geraud hadn't traveled any distance like the others. He lived in Saskatoon and maintained HB's stable of aircraft. During the winter months, he overhauled, painted, and polished every bird so they'd be slick as new by early spring and ready for the grueling summer ahead.

He hadn't toured with the boys in recent years; he was getting too old. Instead, he maintained a liaison with the men on the road and was always a lifesaver when somebody broke a prop, shattered a wing, or needed an expert opinion.

Pop Geraud was the unsung hero of this bunch of sky-happy clowns. He measured success by longevity, and he was far in the lead.

Carter Camp — the greatest of them all, in HB's eyes — was indirectly responsible for this meeting and others past. He gave HB that first taste of flight and changed both their lives from then on.

Carter was somewhere around sixty years old, and he too had lived in Saskatoon ever since that first meeting with HB. Since then, he'd guided HB's flying enterprise with an able assist from "Pop" Geraud, his right-hand man.

It was a good job to grow old in, and, like "Pop", he'd had his day too. He couldn't fly forever; maybe he'd at least die with his boots on.

Time was valuable to HB, and pressing business awaited his attention. He brought the meeting to a close, thanked the oldtimers for showing up another year, and once again welcomed the new faces in the room. He hoped that additional faces would appear tomorrow or the next day, and in the meantime he expected them to spend their time at the field. Under Carter's guidance, they'd ready their planes and equipment for the earliest departure date possible.

He rubbed his pink head as he searched for any forgotten thoughts. His other hand reached for his bottom and scratched vigorously. He quickly withdrew his hands and clamped them together in front of him.

Kid Galahad and Diablo, an odd-ball partnership. Herol the fledgling airman and Tyson the superb pilot, veteran stuntman, harsh taskmaster and best friend ever.
(Author's collection)

8 WHICH WAY IS UP?

Herol sat uncomfortably in the ancient, straightback chair and looked into the rainswept street below. He searched his thoughts for something that he could write to Gramps. Try as he would, there wasn't a thing good to tell, only the sad events of the past twenty four hours. He was reluctant to tell the old man that they were encamped in a seedy "Clean Beds — 50c" hotel. Maybe he'd think of something else eventually.

Tyson was propped up on his elbow in bed and held a small tin that served as an ashtray. He wasn't too careful when he flicked ashes from his cigarette. Some fell in the tin and the remainder scattered on the bedspread. It didn't really matter since the bedspread had seen better days. His wet shoes left a soggy trademark at the foot of the bed, the same place where countless other shoes had been. He'd spoken very little since they entered the room, occasionally humming a tune but silent most of the time. His lifeless eyes stared at the bleak wall across the room, and he fingered his half-burned cigarette near the ancient bed springs. Suddenly he jumped up!

"What you doin'? Ain't the bed OK?" Herol asked.

"I'm lookin' for bedbugs — you better do the same." Tyson had commenced a bug hunt for lack of something better to do.

"If you can't see them danged bugs you can smell 'em. Them critters climb up to the ceiling and drop down on top of you."

Herol tore his bedclothes apart. "I don't see nothin' but I smell kerosene!" Herol said. "How come you told Mr. Bottomlee that we got a sensational act?"

"Thought you'd never ask!"

"I'm sure glad he didn't ask me anything 'cause I couldn't tell him nothin'."

"It's this way. When you get up a routine, at first it's only a general idea, and you sort of fill in the details as you practice it." Tyson sounded rather bored with the explanation.

"I don't know nuthin' about nuthin'."

"Just leave it to me," Tyson assured him.

"How about your partner that got creamed last year? Did he leave it up to you too?"

"Ah! He was knucklehead. He was bound to get creamed."

"How long before I get creamed? I'm goin' home tomorrow! I'd just as soon be a live knucklehead."

"Didn't say you were — leave it to me, remember? For instance, let's take a simple idea like a parachute jump."

"Wow! That ain't simple, I ain't ever jumped!"

"Shut up! Maybe we can dress it up to look more spectacular. We can drop a bundle of rags first, and the suckers think the chute didn't open, or we can spill a sack of flour on the way down, or make a delayed opening close to the ground. That one and the bundle of rags always make the yokels go crazy."

"You got three or four ideas in mind all the time, and the unrehearsed stunts add real zing to a show."

"I don't understand nuthin' you said."

"You will, just as soon as we get a plane from Happy Bottom."

"Good ol' Happy Bottom," said Herol.

"Hey! It's HB to you kid. And if you wanna keep your

job, call him Mr. Bottomlee when he's around. By the way, I'm going out after supper. I got a date with Marge."

"Who's Marge?" Herol asked in surprise.

"HB's secretary, Miss Savageau." He swung his shoes off the bedspread and sat up, giving one of his rare crooked grins.

"When did all this happen?" Herol asked.

"That's one of the stunts you got to learn on your own," Tyson laughed.

Herol sat in the dimlit area downstairs that served as a lobby. It was a lonely night and he'd grown weary looking at the faded wallpaper in his room. His thoughts drifted back and forth from the past to the uncertain future. He looked at the new shoes Dick gave him only minutes before he died. They'd forever remind him of what he was trying so hard to forget.

"To heck with yesterday!" That was Tyson's motto. But how do you forget the past if it was only yesterday! He stood up and made his way back up the stairs to his room.

Live for tomorrow? Wow! I'm only living tonight so that Tyson can get me killed tomorrow!

He unlocked the door, found his bed in the dark, and flopped on it to sleep through a restless night until daybreak.

Sunlight drifted through the dirty window and roused him from his fitful dreams. He was surprised to find himself fully clothed, and then he recalled his troubled evening. He glanced across the room to see Tyson, also clothed, and snoring to his heart's delight.

"Hey! Wake up!"

Herol grabbed Tyson's feet and pulled his legs over the edge of the bed. This roused him enough to stop snoring and he opened his bleary eyes.

"Dang! I feel rough! I killed a whole jug of brew after I took Marge home. She wouldn't drink none of it—she's, got real class. Hey! Where's the washroom?'' he asked as he stumbled out the door.

"IT'S OUT OF ORDER!'' Herol called after him. Then he sat and waited for Tyson to return.

"Hey!'' Tyson complained, "That washroom ain't out of order but it sure smells in there. Some drunk puked all over himself when I walked in. I musta scared the heck outa him! And what do you think he did? He fell into the bathtub and went right to sleep.''

Herol was sick with disgust. Disgust with Tyson, with himself and with the whole stupid world.

Tyson felt Herol's moodiness. "Well what's wrong with you?''

"I'm trying to decide if I wanna live my life like this or not,'' Herol said angrily.

"Oh? Well while you decide, I'm gonna get something to eat. I'll be at the bus depot when you make up your mind.'' And with that Tyson walked out.

Herol brooded for a few minutes and then he jumped up and pounded the door with his fist.

"NUTS! I guess if I'm in, I might as well make the best of it.'' Herol said, slamming the door behind him and bounding the stairs.

For some reason, Tyson wasn't there when he sat down at the depot counter. The waitress brought him a cup of coffee.

Tyson swung his leg over the stool and slid in next to Herol. He was wide awake, well scrubbed and groomed.

"I thought you'd already be here. And how'd you get so neat and shiney looking? You were a mess when you left the room,'' Herol asked, puzzled.

"I cleaned up in the men's room here in the depot — how's the coffee?''

"Coffee's OK. And I'm gonna stick it out—for awhile at least. So what's up for today?''

"I'm glad you're staying,'' Tyson said staring into his coffee cup. "Things'll get better, I promise.''

"As for today," he continued, "we'll see Pop Geraud about the Eaglerock, and we gotta get our parachutes looked over and fitted, and maybe we can take the plane for a short hop so you can try the new bird out. How's that?"

"Sounds great!" Herol laughed.

Pop Geraud gave the biplane a rub with his polishing cloth. Today would be the first time that any of the newly rebuilt planes would be outdoors. All winter the Eaglerock enjoyed his loving care — a freshly overhauled engine,new wooden ribs replaced last summer's fractures, fresh cotton fabric covered the entire structure, and a new paint job topped off his efforts.

He expected that several of the men would be drifting in this morning, and reminded himself that there were other favorites beside the Eaglerock that needed attention this morning. He gave the plane one final rub and turned toward the hanger door just as Bob and Herol came around the corner.

"Hiya Pop. How you doing?" Bob said with genuine warmth in his voice. Pop was held in high esteem by Bob. He appreciated anyone who put their heart into keeping these airplanes flying as well as Pop did.

"Hiya Bob. Good to see you again. Another flying season just about here, right?"

"Yup!" said Bob. "Here, I want you to meet Herol, that is, Kid Galahad. He's my new partner."

"Glad to meet you son. Welcome to the flock," Pop said.

"Thanks," said Herol. "Boy, this sure is a beauty." Herol's eyes roved over the gleaming Eaglerock.

"Do we get this one Pop? Seems like Galahad here, has his heart set on it already."

"She's all yours Bob. She's a real sweetheart. I spent all winter getting her fixed up."

"She's a fine plane," said Herol affectionately as he

wandered into the hangar and looked at the big red and silver biplane. The fresh dope sparkled where the sunshine glanced off her skin. Her long silver wings were eager to embrace the clean air. He touched the lower wing and felt the resilient fabric, tough but soft to the touch, then walked slowly around the craft, hoping that she would be his companion for the months ahead.

"What you think of this turkey?"

Herol's delightful daydreaming halted at the remark. He turned to be confronted by the friendly faces of Tyson and Carter Camp. He didn't know which of them referred to this fine plane as a turkey, so he answered both of them. "I think it's more like a peacock!"

"Welcome to the club, I'm Carter Camp. Guess you're right, Pop changed this tired turkey into a beautiful peacock."

"Happy to meet you."

"After you guys look over the uh — peacock, come into my office." Carter strolled away with a friendly wave of his hand.

Herol marveled how different Tyson seemed now compared to the man he'd known the last couple days. When Dick got creamed, he was cold blooded, he was a slick con artist at the bus depot the other night, and a drunken bum this morning. Now he looked handsome, a real hero type, a strange guy to figure out.

"Hey kid! From now on, you're Kid Galahad."

"How come? Who's that?"

"He was a good guy that run around on a horse and rescued all the fair maidens in distress."

"I ain't never rescued no maidens. I ain't sure I got a girl!"

"I'm talkin' about the act we're gonna put together, and the first act starts today. You're Kid Galahad and I'm Diablo. I been that for a long time. I told Happy Bottom we'd give him a sample today if he can get out here after lunch."

"I don't know what to do. You ain't said nuthin' that I know what you're talkin' about."

"Leave it to me. I got a feel for these situations. We'll hop the lady around the field once so we get to know her. She's anxious to know us too. I'll fit our chutes. If Happy Bottom likes what he sees, I'll get us the fattest bankroll you ever seen. Agreed?"

"Sure, what I got to lose?"

"You might lose your ass! The name of the game starting today is 'You bet your ass', and neither of us want to lose. It's all business from now on. Let's take the bird for a hop."

Herol stepped lightly on the wingwalk and flung his leg into the rear cockpit and settled himself in behind the windshield. Tyson stood in front of the big wooden propeller, its scimitar blades flashing like two wooden knives in the sun.

"Contact!" Tyson gave a hefty swing, the engine caught hold and the familiar bark from the short exhaust stacks sang a throaty tune. He kicked the chocks from under the wheels and clambered into the front cockpit.

"How about helmet and goggles?" Herol asked.

"To hell with them," Tyson hollered.

Clouds of dust and dry grass swirled up and around the cockpit as the tail skid dug a furrow behind them. They taxied downwind toward a wire fence that held back a herd of curious heifers. Herol gunned the engine and swung the Eaglerock into a turn, the outer wings extending close enough to unnerve a few of the more timid creatures and send them galloping off with their tails flying. He laughed and headed the ship into the wind and pushed the throttle home.

The majestic craft gained momentum and then leaped ahead of the dust storm that swirled around the propeller. They lifted into clean air on easy wings, the engine droned contentedly as it coaxed them ever higher into the heavens. The propeller turned a pirouette, transforming it into a halo that engulfed the sky. Perhaps some of the craft's exhilaration rubbed off on them, because they too felt the surge of excitement as they sped from the realm of earthbound humanity toward the infinite heights.

Herol wished they'd climb to heaven and beyond. The white pillows floating past his wings made him forget once more the world he'd escaped from.

Tyson turned halfway around in the front cockpit and banged on the cowling. Herol was caught daydreaming and knew it! He'd climbed several hundred feet and didn't have the sense to level off, and Tyson's intent was a quick hop around the field. The stick shook as Tyson took over control in the front cockpit and eased the plane back toward the field and in a shallow glide for landing. He shook the stick again, with "You got it." Herol had it made now. The wheels would probably clear the fence by ten feet and scare those crazy heifers into a hightail run again.

It puzzled him to see Tyson shaking his head —

What's wrong now he wondered? And then it was obvious to him. They weren't about to land over the fence like he wanted. They'd be lucky if they landed in the next field! A hard tap on the stick again, Bob's signal to let him take over.

Tyson eased into a steep forward slip then pulled the nose up at an acute angle to slow the plane. Then he quickly dropped the nose down to a normal glide. "You got it!" The Eaglerock must have sensed Herol's frustration and, being the gentle lady that she was, settled lightly into the grass. Tyson turned around and flashed a crooked grin, then he pointed toward the hangar.

They sat in the cockpit a few minutes while the engine idled and cooled down. Finally Tyson cut the ignition and stepped out as the propeller ticked over to a stop.

"It was a good flight, I got a couple things to talk about before Happy Bottom shows up. Let's sit under the wing and talk, nobody will bother us there. You got some fast schoolhousin' to learn. They ducked under the wings and found comfortable places to sit. Tyson propped himself against the sunny side of a wheel and Herol sat closeby, his knees pulled under his chin and his hands clasped together around his legs.

"I sure did some rotten flying," Herol opened the conversation.

"I said it was a good flight, didn't I?" Tyson countered with a flat voice and blank expression.

Herol looked at his companion, a total stranger from the masterful pilot of minutes ago.

"Like I said, you got some more schoolhousin' to learn. Don't worry about the flyin', you're OK, just need more experience. There's a few pointers you gotta get wised up on. Have you ever wondered why a plane is always a 'she' and never a 'he'?

"Guess I never thought about it."

"Airplanes are ladies — so you always treat 'em like ladies. Treat 'em gentle always, love 'em, and they'll gentle you and take good care of you. Don't ever try to force 'em. Persuade 'em, but always gentle. They get ornery as hell when you get impatient, and that's when you lose the game we start playing today."

"What game?"

"You forgot already? *You bet your ass!* Remember? You only lose once."

"Now I'm talkin' ladies, not broads, there's a difference. Ladies is special, broads is tramps."

Herol felt the blood surging through his veins as he tried to smother the blush coming over his face. This strange flying lesson promised to be most unusual.

"Now you treat 'em different, the broads. They sorta remind me of a danged streetcar. You pay your money, they're off with a jerk, and you ain't figured out what in hell happened. You understand me?"

"I don't know." Herol wasn't sure about anything anymore.

"It's like that dumb takeoff and that approach you made. You didn't appreciate the fine lady you were with. You was taking a joyride on a danged streetcar!"

Herol gulped and feigned understanding.

"You had a lady, and you didn't gentle her. You tried to force her and she got ornery, and you got all screwed up. From now on, I don't want to feel those controls move, only gentle pressure. Caress the lady because you love her and she'll respond."

He didn't blush this time but comprehended what Tyson was trying to say. His peculiar philosophy had a virtue all its own.

"Eventually, you'll find that *you're* flying, and you're taking your lady along for a ride. She depends on you and trusts you, so don't let her down. Understand?"

"You know the difference between a forward slip and a side slip? Which one did I do when I got you straightened out for that landing?

"Well uh — "

"That was a forward slip. It's safer close to the ground. Your nose ain't cocked way up towards Maggie's drawers but aimed ahead in your landing direction, and them exhaust valves don't get a cold blast of air up them stacks — understand?"

"I think so." The odd wisdom made sense, although he'd never thought of Maggie's drawers being part of flying. He wondered if Maggie was supposed to be a lady or one of them other kind. He decided not to ask.

"When we was flyin' I got to know the lady, and she got to know us. We like each other, and she trusted us. What did you do?"

Herol looked at his companion with a mixture of admiration and dread. "I—uh—did the same thing."

"You didn't do a damned thing!" His bleak eyes gave Herol a look that penetrated his soul. He wanted to run but couldn't move. Those profane expressions weren't ever meant to be funny but another facet of this strange man.

The uncomplimentary retort stung Herol, and he found himself doing the same as Tyson. They both sat motionless and stared at the ground. When he finally looked up, his partner was a different man — almost handsome — the tortured eyes were gentle.

"Enough flyin' lesson for the first day. You ever made a jump?"

"Me? No!"

"Doesn't matter, sooner or later. We'll do something for Happy Bottom today and tell him it's only a warmup and not the real thing."

A plan jelled in Tyson's mind, one that would instill confidence in the boy. It could turn out to be pretty hairy if the kid went to pieces, but that's what he had to find out. He needed a solid clue as to whether he should encourage this partnership or send him back home before it was too late.

"That sounds pretty good to me." Herol felt relief that he wouldn't have to jump.

"Carter asked us to stop by. Maybe we can scrounge a cup of coffee."

They walked towards the hangar, Tyson was lost in thought and Herol much subdued from the crude philosophy of his flying lesson. The hangar doors were partially open, revealing an orderly interior. Sunlight drifted through the windows along the wall and filtered through two large skylights in the roof. The rear wall was partitioned off for a workshop and a small office that Carter and Pop shared.

Carter was wading through a sheaf of invoices while Pop busied himself in the corner of the hangar wiping spilled oil from the floor.

"Come in! Be with you in a minute. Have some coffee. It's always hot — over there behind the file cabinet. Your chutes are on the bench next to it. I got the receipts here someplace for you to sign."

"Yeah, thanks, we'll try the coffee." Tyson reached for

the stack of paper cups on top the cabinet and handed one to Herol.

"Any word from HB if he's comin' out this afternoon," Tyson asked.

"Pretty soon I guess." Carter scanned the last of his bills and tossed them in the bottom drawer. "He called this morning and said he and Miss Savageau would leave the bank around eleven for lunch, then they'd drive out here."

"What in heck is Happy Bottom doin' with Miss Savageau?"

"Don't know, and I ain't gonna ask."

"Let's get the heck out of here!" Tyson threw his half filled cup into the trash can."

"No reason to get mad about it. Don't forget to sign for your chutes."

"Sorry, I must be in the spring-loaded position this morning." Tyson replied. "If they — uh — Happy Bottom shows up, give a holler. We'll be ready to leap off and give him a show." They picked up their chutes and left.

"If you're gonna be Kid Galahad, you better start acting like a pro, or at least look like you know what you're doin'. Learn how to carry your chute, don't bundle it under your arm like a sack of crap. Do like this. Pick it up by the leg straps and carry it over your shoulder, or better still, put the harness on and fold the pack over your fanny. It's your chute and your life that you're takin' care of."

Herol dutifully swung the harness across his shoulders and tucked the parachute over his fanny.

Tyson adjusted their chutes with the usual amount of colorful wisdom. He showed how easy it was to fall out of a harness and hang upside down, or worse yet, fall completely out. The leg straps were important. Loose and comfortable "and you'll hurt yourself when the chute opens and you'll be screaming in a high soprano all the way down. They gotta be tight! — like so!" A few more sage remarks ended the session for the time being.

Tyson's brain spun at top speed. His formula had to be worked out quickly, something spectacular looking for Happy Bottom yet reasonably safe for the kid. If the kid panicked, all hell could break loose. He'd implied they wouldn't jump, but after a primary lesson in wingwalking maybe he'd change his mind if the kid shaped up.

"Like I said, we'll do some routine stunts, and after we get HB's attention, we'll teach you a bit of wingwalking."

"Gee! What I gotta do?" It was now or never, and Herol sincerely wished it was never.

"Ah shoot, it's a cinch. Everybody watchin' is more interested in if you're gonna get killed then in what you're tryin' to do. The first thing you do before you get out of the cockpit is to button up tight. Don't give the wind a chance to open you up and knock you off balance. And when you're ready to crawl out, take it slow and hang on to everything in sight. Watch me for signals, cause I'll be flyin' to help you along. You can wear my heavy gloves, easier to hold on to the wires. Think you're game?"

"Never know if I don't try."

"Remember, we play it by ear, and I'll do all the thinking for you. You gotta trust me, even if you don't, understand?"

"Sure," he said. But he didn't.

"Where's your helmet and goggles? Take mine, you'll need 'em worse than me. And where's your chute?"

"The front cockpit."

"Switch 'em around, you might as well make the takeoff. We might change seats in the air, all depends on how we shape up. Startin' today, the flyin' is gonna get rough, and you'll pee your pants a few times before you learn the ropes. You'll wish I was dead, and that might happen someday too."

"Why did he have to say that. I'm about to pee my pants now," Herol mumbled to himself.

"What say?"

"Nothin', just thinkin'."

“They’re driving in now!” Carter yelled between cupped hands.

“Thanks, we’ll head for the blue,” Tyson yelled back.

An immaculate black limousine rolled slowly to a stop near the corner of the hangar. Tyson stood by the plane and looked in the direction of the car, trying to distinguish the occupants. HB stepped jauntily out and walked around the front end to the opposite door. He opened it, stepped aside, and Miss Savageau appeared. HB waved, and when Tyson returned the greeting, he thought that Marge’s right hand made a furtive movement of recognition.

“Ready?” Tyson called.

“Ready!”

“Contact!”

The engine caught and roared.

“Let’s go.” He threw his legs into the front cockpit and motioned forward as Herol gave her the gun.

Tyson tried to concentrate on the minutes ahead but kept seeing Marge with Happy Bottom. “Even ladies is tramps to go out with a bum like me. To hell with all of ’em. I gotta think about not makin’ this hop a disaster.”

They climbed in lazy circles, keeping directly in view of HB’s limousine. His course of action was falling in place. Part one, he’d demonstrate a short session of wingwalking to give the kid a shot of courage. Parts two and three would depend on Herol’s first reactions.

“Safety belt on tight?” Tyson rolled the long silver wings gracefully around the horizon. Herol banged his head against the side of the cockpit before he realized the wings were level — except they were upside down! By the time he collected his wits, he felt Tyson’s smooth pressure on the controls and the nose plummeted toward the brown patchwork of fields below. Gravity pushed him hard into his seat and the wind screamed past the windshield nearly drowning out the wild howl of the flying wires. The feisty lady skimmed over the hangar roof and was hell-bent for the sun. Tyson nursed the last bit of lift from the wings as the nose passed vertical and rolled slowly back to level flight. He turned around in the cockpit and gave Herol a reassuring grin. The Eaglerock was a ballerina in Tyson's hands, and Tyson was the dancemaster.

"Isn't he wonderful!" Miss Savageau exclaimed.

"I haven't seen a thing yet; he's supposed to come up with something new." HB gave her a sidelong glance.

Tyson smiled to himself when he saw Marge waving. He positioned the craft so the spectators would have a grandstand seat from the ground. He pulled the throttle back and yelled, "Take over! Stay in the same area, fly smooth."

He unbuckled his chute and safety belt, grasped the left center section strut, and hoisted himself half out of the cockpit, then turned around to watch the kid's reaction — so far so good. He lifted one leg out of the cockpit and lowered himself to the wingwalk, grasped the cockpit edge and planted his other foot down. He wished he hadn't given the kid his goggles and gloves, but no matter, he'd been through this hundreds of times before. He'd play it extra safe today. Then he was gone.

"Where did he go?" Herol felt a slight shift in the center of gravity. The bird was nose heavy. He peered over one side of the cockpit and then the other. He must be seeing things! He looked again. Sure enough, Tyson dangled by one foot from a cable attached to the landing gear. He looked again and couldn't find him. "He's nuts, or I am!" Tyson was swinging from one side of the nose to the other, scant inches away from the propeller! Occasionally he got a glimpse of him, and then Tyson disappeared for what seemed hours, eventually to reappear flat on his belly across the leading edge of the lower wing, his legs flailing the air for balance and his feet missing that wooden meat grinder by a shadow. He made a lunge and grabbed a wire and pulled his knees up and over the wing.

Herol breathed a sigh of relief, happy to see him relatively safe again. He hoped the show was over. But Tyson had different ideas. He half flew, half walked outboard toward the wingtips. Aileron control felt heavy with Tyson's weight, and they started to turn in a shallow bank. Like a spider in a web, Tyson scrambled uphill on the wires until Herol had enough control to level the wings again.

"If the kid only knew it, I'm even flying the damned plane for him!" He glanced briefly at the ground. "We're still in perfect position!"

The wind tore at Tyson's bare head and half-closed eyes. A perpetual grin never left his face, the grin that kept him sane in this insane world of his. He reached the cockpit, leaped in, then out again on the opposite side. His arms and legs performed a fantastic dance as he sped through the maze of wires. The plane heeled over into a steep bank, and once more he clambered uphill in his precarious web.

Again he dove into the cockpit, to reappear on his way up and over the center section, then he jumped down to straddle the engine cowl. His legs were bent hard back at the knees while his feet held fast to the forward center section struts.

"These exhaust stacks are a pain in the ass — the feet too!" He always talked to himself during these sessions, especially if he hadn't had a drink to fortify his courage. Being sober was sheer terror. "Ah, the hell with it!"

Herol ceased to be surprised at anything. This was a crazy day from the start, and it wasn't over yet. He continued to 'fly smooth' as he'd been told.

The propeller blast was a tornado tearing through Tyson's hair and threatening to gouge out his eyes. "Damn! Wish I had a drink, and them damned goggles." He uncocked his feet from the struts, pulled himself to a kneeling position, then slid backward until his feet touched the rim of the front cockpit.

"Enough of this crap! I gotta get the kid started before he's got time to think." He settled himself in the cockpit, sat for a moment to catch his breath, then buckled himself in. He turned around, grinned, and shouted, "You wanna try?"

Herol gulped. He knew it had to happen sometime, but couldn't believe it was now.

"Pull yourself into the front cockpit — headfirst," Tyson shouted.

The engine was ticking over slowly, just enough power to maintain altitude. He hoped the first exposure to the propeller blast wouldn't panic the kid. He turned around again to see him halfway out and gripping the padded cowl in front of him. He nodded encouragingly and motioned, "Come on over the top."

Herol gave a sudden lunge and landed headfirst on the floor, his legs gyrating in the slipstream. He crammed himself in, got untangled, and took off his chute. He puffed from a combination of fright, excitement, and exhaustion. Tyson grinned, Herol laughed — the ice was broken!

No point in letting him think. Tyson pointed to the strut and gave him a gentle nudge. "Climb out like you saw me do. Be sure that you're buttoned up tight, stay on the catwalk over the spar or you'll punch through the fabric — and don't look down! See if you can make it out to the end of the wing and back."

Herol nodded dumbly and hoped that Lady Luck would take care of another damned fool. His stepfather, Mr. Daley, was right after all! He gave one last look at the cockpit and pulled himself out like he saw Tyson do.

Tyson watched for signs of panic, ready to call it off on a moment's notice. The kid was scared stiff, but he'd be all right if he took it slow and easy. The next few minutes would tell if the plan he had in mind would materialize. He'd help the kid along — a slight skid to help him reach a wire, or slow the bird down to give him needed momentum, or a gentle bank to encourage him back to the cockpit. He'd wait and see if the second and third parts of his scheme would be realized.

Herol's back was to the cockpit as he progressed cautiously out the wing. His gloved hand was wrapped around the forward landing wire while his feet were jammed in front of the flying and drag wires. He felt reasonably secure, and the wind didn't seem so bad once he moved outboard of the propeller blast. His hands shook with the wires, vibration or fright, he wasn't sure. His feet beat a tatoo on the wings as he attempted to change position. He looked over his shoulder for Tyson and felt comfort when he was given a reassuring nod.

He stepped away from his wire braced foothold, then teetered as a gust of wind caught him off balance. His foot slipped and his fingers instinctively froze to the wires. A painful sting shot across his back as he was buffeted against the underside of the rear flying wires. He caught his breath, afraid to think, and fought the sick feeling in his gut as his legs trailed behind him. Cold sweat trickled down his neck and he fought the panic that threatened. He looked toward the cockpit and was reassured by Tyson's grin. Tyson cracked the throttle back to idle and Herol's flailing legs banged against the trailing edge of the wing as they decelerated. He scrambled back to security and planted his feet down on the wing again.

Tyson reminded himself that this little incident looked pretty good from the ground. Audiences always thirsted for blood, even their friend's . He rocked the wings gently to attract Herol's attention and beckoned him back to the cockpit. The return was negotiated with a surer step. The cockpit meant safety. Tyson still held the engine near idle when Herol stuck his head in the cockpit.

"Think you can do it again — all the way to the tips this time?"

"Hell yes! I'm a stuntman, ain't I?"

They'd descended to about five hundred feet, just right

for the next turn of events. So far the scheme was a success. Act one was complete although Herol didn't know it. Act two coming up! Tyson watched the youth making confident progress toward the interplane struts, and when he had a firm grasp of them, Tyson released aileron pressure and allowed the left wing to sink.

"Hey! What's happened?"

Act two was in progress! The engine sputtered, backfired, then quit! The propeller windmilled a few turns and stopped. Herol's eyes were glued on Tyson who looked back with a resigned expression. The soft whisper of wires was the only sound that broke the silence.

"I'm goin' in for a dead stick landing. Hang on!"

Herol's weight near the wingtip left little choice but to continue in the spiral, exactly the way Tyson planned it! If Lady Luck smiled kindly, they'd land within fifty feet of Happy Bottom's limousine. It could get hairy. They'd need extra speed and would have to land crosswind to help support the kid's weight.

"You okay?" Tyson shouted, aware that their voices carried easily to the ground.

"Yeah, guess so, if I don't pee my pants on the way down!"

"Nothin' will happen." Tyson needed to keep talking so the kid would forget to be scared.

"Wanna bet?" Herol mumbled.

Tyson reminded himself to caution Herol about sound carrying to the ground. This talking stuff was great provided it was rehearsed — no cussing! He got back to the serious business of coaxing this lady around the last quarter of her spiral. They turned crosswind, settling slowly and lined up with the ramp in front of the hangar — right in the lap of Happy Bottom and company!

Herol alternated between fits of panic and hope as he watched the ground rushing toward them, while Tyson coaxed the feisty lady to try a little harder.

They were settling in, inches above the grass, Herol's

weight holding the left wing low against the crosswind. The left wheel jolted against the bungee cord and the tail skid dug in simultaneously. The right wheel touched down, then lifted out of the grass and remained poised in midair as the bird fought against the tailskid, trying to weathercock into the wind.

"Those crazy nuts are going to hit us! Everybody run!" HB screamed as he waddled off as fast as his fat legs would carry him.

Miss Savageau turned, then looked back at the approaching craft and decided to stay where she was. She didn't feel any sense of danger. Bob was flying — that was reason enough for her.

Carter and Pop still leaned against the half-open hangar doors, the same position they'd assumed when the Eaglerock took off. The grandstand show wasn't over yet and they weren't about to miss the finale!

Tyson ruddered into a slow ground loop to downwind position after Herol scrambled inboard toward the front cockpit. The show wasn't over yet. Act three coming up — the tough one. He was pleased to see Marge standing close enough to touch their wingtip, her radiant smile speaking a thousand words. He nodded to Carter and Pop and saved his best smile for the girl.

Herol hoped that Mr. Bottomlee noticed him, and the others too! He waved to his audience like a conquering hero. He'd caught the applause fever and hadn't heard a shout or handclap yet!

"Cut out the damned grandstandin' and pull the prop through!" Tyson jolted him out of his glory dream.

"I thought it broke?" Herol said.

"Don't ask questions! Prop it!"

"She's pi — uh — getting gas. Contact!" Herol shouted.

The Hisso belched smoke, then settled down to a contented roar.

"Get in!" Tyson shouted over his shoulder.

"My chute's in front with you."

"Get in! You can pick it up later."

They were hightailing downwind while Herol struggled into the rear cockpit. The bucket seat was too low without his chute so he decided to stand up.

Tyson was thumping on the cowling before the wheels lifted out of the grass.

"Hey! Come get your chute."

"We're still on the ground. What's the hurry? he grumbled as Tyson waved impatiently.

"This is crazy!" He started to climb out. "First time I did this was to get rid of my chute, now I gotta go back after it. What we wear 'em for anyway?"

So far so good. Tyson helped untangle his companion in the bottom of the cockpit. He glanced over the side, still in good position and climbing through 200 feet.

Herol busied himself fastening his parachute harness when Tyson motioned him to return to the rear cockpit.

"Get your feet out first, straddle the rear windshield and hold on to the rear of my cockpit, then just back in. Get goin'!"

"This is easier said than done," he muttered.

"I'm in a hurry!" yelled Tyson.

It was important to keep Herol off balance, keep him confused, if act three was to come off. Timing had to be perfect. He watched him struggle over the cowling, then coaxed a little more altitude from the bird. A gentle turn brought them in line with the hangar. Now was the time, the kid was just sitting down.

The nose of the Eaglerock ducked below the horizon and they dove until flying wires screamed in protest. He cast a sidelong glance to the rear, and, as hoped, Herol's gloved hands gripped the cockpit. He was immobilized.

They hurtled toward the ground until the proud Eaglerock thought they were doomed. Tyson eased the nose up in time to thunder over the hangar roof once more and zoomed into a graceful loop. This was the finale, and it had to be right. He cast another glance to the rear. Herol was bug-eyed and hanging on for dear life. If his hunch was right, the kid didn't have time to buckle his safety belt. Tyson threw his head back to find the horizon as the Eaglerock's nose curved over the top. He relaxed the pressure on the elevators, and dry grass and dirt sprinkled off the floor as the loop flattened out at the top. Sure enough! A pair of gyrating arms and legs scrambled out of the rear cockpit!

"Hey! What's going on!"

Herol tumbled end over end. Tyson's goggles whipped off his eyes and beat a tatoo on the back of his head. He reached for the retaining straps and tore them loose. As he tumbled over he caught a glimpse of the hangar between his legs and the goggles floating beside him. The ground was rushing toward him at a frightening speed.

Tyson circled in a lazy turn waiting for the white canopy to mushroom. He grinned with anticipation at the expected tongue lashing he'd get from the kid, then peered over the side to see how his "stuntman" was progressing.

"My God! He's still falling!" He corkscrewed on a wing and dove for the ground. "If that chute don't open in the next few seconds, he'll be a bag of guts!"

"PULL THE RIPCORD KID!" Tyson yelled.

WHUMP!!! The chute opened just as Herol was in the middle of a midair somersault. The sudden shock snapped his whole body like a wet towel knocking the breath out of him. In pain he reached up and grabbed the risers. Then he looked down just in time to see his impending collision with the ground — whereupon he fainted.

The violent impact with the earth sent spears of agony through his body and brought him abrubtly back to giddy consciousness. The canopy billowed and raced across the grass toward the hangar, dragging Herol's pained and bruised form along behind it.

Grey shadows weaved in front of his eyes, growing

more distinct when his fuzzy brain cleared. As if still in a dream, there was Tyson grinning from the cockpit as he taxied alongside the runaway chute.

"Spill your chute!" he shouted over the engine.

"Ouch!" Herol groaned as his scrawny bottom slid across a patch of sharp rocks.

"Spill your danged chute, you'll wind up inside the hangar!"

"Oh yeah." He grabbed the risers and tugged at the lanyards until the canopy collapsed in front of the hangar.

Tyson was out of the plane before the propeller unwound, helping him to get up on wobbly legs. They grinned, both proud of each other.

"This flyin' might beat heck out of workin' but it's sure a pain in the fanny." Herol laughed and explored his shredded backside.

"You're a crummy looking Galahad! Old King Arthur would fall off his horse if he saw you."

"Who's King Arthur?"

"He's the guy who 'entertained' all them fair maidens that Sir Galahad rescued!"

"He did?" said Herol, still not really understanding.

Well, anyway, he guessed he had plenty to write and tell Gramps about now.

A J-1 'Standard' biplane similar to the planes in which Johnnie Day and 'Black Jack' Reeves performed their aerial magic. In the background is a Ford Tri-motor.
(Author's collection)

9 COOL DAYS, HOT PILOTS

Tyson was awake at daybreak. He pulled himself up on an elbow and looked across the room at his companion who lay sprawled on his stomach beneath a pile of blankets.

"I'll bet he's sure beat up after that hard drop yesterday."

He lifted his legs off the bed and lowered his feet to the floor, pulled a blanket around his shoulders, and reached for a cigarette. Me and the kid got to do some talkin', he thought. He's gotta be able to take care of himself if I get creamed one of these days.

"Get up!"

"Get up!" Tyson growled again.

"What time is it?" Herol moaned.

"I don't know but the sun's up." Tyson pulled a shirt over his shoulders and reached for his socks.

Herol eased out of bed, touching his aching muscles, and felt the bruises that seemed to be everywhere. He tried to open an eyelid, then gave up when he felt the swollen eye.

"You sure got a beautiful mouse on that eye, the other one looks almost as pretty."

Tyson dressed, rubbed the sleep out of his eyes, slicked his hair down, and stuffed his razor in a coat pocket.

"Hurry up! Let's get some breakfast," he scolded.

"Yeah, just a minute, You sure did some sneaky things yesterday. You said we'd play it by ear. You should have said fanny — you almost made me bust mine."

"What you griping about? Happy Bottom got you figured for a red-hot stuntman! Hurry up!" He opened the door and left.

Herol mumbled as he pulled his clothes on. He vented his wrath on Tyson, only to be reminded that through him he was on the road to glory!

"What's good?" Herol asked sitting down next to Tyson at the counter.

"The coffee is better today," Tyson said.

"Gimme some coffee and toast," Herol asked as the waitress walked up.

"One burnt stack," she cried to the unseen cook.

"We got business with Happy Bottom today. We'll nail him to the wall if I get to him before he cools off. I'm sure he liked that routine we gave him yesterday."

"Was that part of the show?" Herol was bugeyed.

"Sure, why not?"

"I'll get killed if we try that again."

"No you won't. Anyway it's gonna get rougher every day. You wanna back out?" Tyson asked.

"No, but you gotta help me out so I don't get creamed too soon — Okay?"

"Yeah, trust me. I'll bet you didn't even wash your face this morning."

"My face! I can't wash nothin' for a year! I'm sore all over!" Herol protested.

"Wanna come along when I give Happy Bottom the shaft this morning?"

"Nah, I look too beat up. He might figure I won't last the season. You think I will?"

"You'll be okay — get yourself patched up. Carter will have lots of iodine and bandages out at the field."

"Will you be coming out after you see Happy Bottom?"

"I'll be there. Trust me to make a deal without you?"

"After yesterday, I don't trust nobody — especially you!" Herol growled. Then he smiled, "Yeah, okay," and drank his coffee.

The morning promised a crisp breezy day. Early spring was still hanging on to the memories of winter. Patches of frost painted the few blades of grass and spring flowers that peeped from the ground. It was an ideal day for the airmen to unwind and flex their egos. Some might fly, others would tinker with the planes, and a few would swap yarns with newfound friends.

Herol kicked at loose gravel on the road as he walked briskly along. He could see the rows of wooden hangars from the road and the sound of intermittent coughing from cold engines filled the air. He followed the curve in the road that led to the main hangar, and as he approached the oil and cinder ramp he searched for the Eaglerock that was nesting inside.

Several men stood in a group. An occasional solitary figure leaned against a hangar or walked between the array of planes aligned next to the road. He didn't recognize anyone. The leather boots, jackets, and helmets made them all look too much alike. He'd single out a loner, or, maybe Carter or Pop would introduce him around.

His path led to the hangar where the feisty Eaglerock perched inside the door. He cast an admiring glance at her massive silver wings, hoping that some day he might own such a beautiful bird.

"Nice lookin' boid, ain't it?" chirped a goggle-eyed stranger in a strange sounding accent.

"Sure is! My partner and I have her leased for the season."

"Dat's great! I'm Bernie — I'm a stuntman in dis outfit. Who you?"

"I'm Herol, nice to meet you Bernie."

"I ain't got no partner, don't need one. I ain't had a chance to meet any of dese bums yet. Where youse get da shiners? Musta been out wid a fancy bimbo!"

Herol thought this guy must be a real squirrel, or else he's putting on a terrific act. "I got banged up yesterday when we was practicing. You got an act?"

"Yeah! I got a million of 'em — all kinds of jumps and tricks. You name it, I'll do it!" Bernie was getting excited over his make-believe prowess.

"If you don't have a partner, maybe you could join up with us. Bob Tyson is my partner — he'd have to okay it, I guess." Herol regretted his rash offer and wondered how he could back out of it.

"Yeah, dat would be da nuts." Bernie sounded less enthusiastic than before.

The roar of engines thundered across the field, and everyone instinctively ducked. The sky filled with fabric wings as two red and silver Standards in close formation shaved the hangar roof and peeled off in opposite directions to land. The one on the right made a spectacular sideslip to a landing, while the other craft wallowed out of a chandelle and hesitated for a split second as if uncertain what to do next.

"BANG!"

A sharp crack like a rifle shot rent the air, followed by a sudden burst of power from the engine.

"HE'S BUSTED SOMETHIN'!" one of the leather-clad group shouted.

"That's Johnnie Day. I'd recognize that heap of rags and baling wire anyplace," one of them shouted.

Black Jack clambered from the cockpit of his Standard, his eyes riveted on the stricken plane of his comrade as it gyrated through the sky like a crippled bird.

"How about one of you hot pilots get a car — we gotta get to him!" His huge frame towered over them and his gruff voice stirred them to action.

Johnnie Day wrestled with the plane and gave her enough rein to pick up flying speed. Crippled wings groaned and shifted as flailing wires and disjointed struts rapped against the wings and cut neat incisions in the fabric. He shut off the ignition and fuel, anticipating a crash.

The bird gave a sigh as she heaved her head skyward and broken wings dropped to near vertical as the ground rushed up to claim its trophy. He braced himself for the impact, his vision blurred as earth and sky merged into a whirling kaleidoscope of color. The airplane hit the ground and Johnnie Day let out an agonized scream as if a white hot poker had seared through his guts.

Black Jack shouted obscenities to the wind as he stood in the back of Pop's Model A truck. They bounced from one rut to the next across the stubble field, the luckless driver wincing each time Black Jack's fist crashed on the roof of the truck. A sprawling mass of arms and legs clung to the sides and tailgate beneath Black Jack's massive boots.

A dust cloud settled around the shallow hole the engine had chewed into the ground. Eddies of hot steam hissed from ruptured water lines, and cylinders crackled as dissipating heat rose into the frosty air. The wings were telescoped into a mountain of wooden spars and ribs. The fuselage rested on its side, broken in two at the rear cockpit. Johnnie was hunched over, half standing, half sitting, outside the demolished cockpit. His face was the color of clay. A weak grin tried to mask the predicament he was in.

Black Jack stampeded over heads, bellies, legs, and arms when he leaped from the truck and ran toward his friend. The rest of them unscrambled their limbs and followed in close pursuit.

"Don't touch me!" Johnnie tried to grin.

"You'll be okay — got an ambulance on the way. They'll fix you up," Black Jack said in a half whisper.

The others stood in a semi-circle behind them at a respectful distance. Someone coughed and gagged, then there was silence except for the gurgling sound that came with Johnnie's breathing.

"Good show while it lasted," Johnnie agonized.

"Sure! Can I help you get untangled from this mess?"

"For Christ sake! Don't touch me! I need some water," he gurgled, and wiped the red foam off his lips.

Black Jack half spun on a heel to ask an unspoken question to his companions. Futile gestures answered him. One of them looked half heartedly in the truck and replied the same as the others.

"The guy in the ambulance will have a drink." He looked across the field and heard the distant whine of a siren.

"My guts are on fire!" He raised his blood-stained hand from his leather jacket and gripped the edge of the cockpit, then surveyed the small audience with a detached look. He moved ever so little, his eyes glazed with pain, then he fainted.

Black Jack followed the trail of blood from Johnnie's boots up to his jacket, and lifted it, revealing the reason for Johnnie's curious stance. He was impaled on a wooden longeron. Its splintered tip had pierced his ribs and emerged between his shoulder blades, giving him a grotesque humpback appearance.

The ambulance raced in front of a dust cloud while it careened over the rough ground and screeched to a halt. A white clad figure ran toward the scene.

Black Jack's scowling contenance betrayed the compassion for his friend as he opened his eyes.

"Give me some water!"

The ambulance attendant deftly reached for the stricken pilot's arm and with a needle injected a drug to ease the pain.

"Gimme some water — please."

The attendant shook his head 'no', changed his mind,

then turned to the driver and motioned him to bring some.

"Sure, right away!"

"Thanks," Johnnie whispered as the attendant busied himself giving what comfort he could.

"Where's that goddam water!" Black Jack stormed as he lumbered toward the ambulance.

"It's stuck in this compartment — can't get it loose."

"Tear it apart, do something! Get it over there quick, or I'll tear you apart." He growled, then wheeled back toward his friend. "It'll be here in a second."

"I hope so!" He spit blood through his lips. "I don't hurt much now — hardly feel a thing."

Herol stood in the background with the others. He'd never met Johnnie but heard about him. The airman's bond had made them friends long before they met. Herol was reminded of similar tragedies that were etched in his mind and wished that Tyson was here. His impersonal attitude was needed by all of them to face the shock of reality. He inched forward, ashamed of his curiosity, but compelled to meet Johnnie, even if it might be the only time.

"Water!" Johnnie pleaded.

Black Jack stomped over to the ambulance, shoved the driver aside, and tore savagely at the metal compartment with his hands.

"Some guy — can't remember — said 'I live to fly, I don't fly to live.' Who said that?"

Herol stepped a pace closer. "Kiddie Karr said that — saw him last week in Fargo."

"Yeah — Kiddie Karr. Tell him I got a new twist on his favorite saying. 'I die to fly! I don't fly to die!' How's that for laughs? — Who are you?"

"I'm Herol, Tyson's new partner."

"Yeah, Happy Bottom said you guys was here. Nice to meet you. I'd shake hands, but I can't let go. Say hi to Bob. Tell him to 'count the sunsets' — don't forget!"

Black Jack's efforts finally paid off. He struggled with a large container and tried to pour water into a paper cup.

The weakening cries of his friend made him all thumbs. Johnnie smiled when Black Jack offered him the cup in his outstretched hand.

"Here you are, buddy." Black Jack raised the cup to his friend's lips. The water dribbled into his mouth and flowed down his chin. Vacant eyes stared into eternity. Johnnie wasn't thirsty anymore.

"Sacre bleu! You sons of bitches! You couldn't give a dying man a drink of water!" Black Jack's arms flailed the air in a hopeless gesture as he lunged toward the nearest body that could fall under his blows. He stumbled blindly ahead, uncertain what to do next. He slumped down in a heap and pounded the ground with his fists. The big man wept.

Tyson slumped comfortably in a leather chair outside of HB's office while Miss Savageau tried to appear busy, aware that his pale eyes followed her every move.

Her thoughts of tender moments shared with him were mixed with an unknown dread. How could he be so gentle and frightening at the same time? He returned her glance with a smile that masked his inner conflict.

"Cold in here!" Miss Savageau shuddered at the sudden chill.

"Guess I didn't notice." Tyson's knuckles showed white as he clutched the armrests of the chair, his face the color of parchment.

"Yeah. Count the sunsets."

"What's that, Bob?" She looked up from her typewriter.

"Nothin', musta been thinkin' out loud."

"Sorry, we'll have to cancel our appointment," HB said waddling through the door. "There's been an accident, perhaps we can talk in the car. Care to ride along with Miss Savageau and I to the field?" He cast a furtive glance in her direction.

"Who got creamed-uh-hurt?"

"Either Black Jack or Johnnie Day. A Standard crashed near the field, two planes were raising Cain. This is bad publicity — our first day, and we haven't even started."

HB saw Tyson's fists clench, the less said the better.

The Model A chugged slowly back to the hangar, its occupants silent. Black Jack remained with the ambulance crew to share the grim task at hand.

Tyson and Miss Savageau stood beside the limousine waiting for Happy Bottom who'd scampered off to question the men climbing out of the truck.

"Hey! Did he get creamed?" Bernie asked excitedly.

"He's dead," Herol answered without looking.

"Who?" The goggle eyes grew larger.

"Johnnie Day."

Bernie's excitement soared to new dimensions as he collared Herol.

"Bet it was hairy, blood 'n guts, everything!"

Herol squirmed away and quickened his step while Bernie two-stepped to keep up the pace.

"Why don't you ask Black Jack about it? He's out there with the ambulance crew."

"Yeah, da meat wagon!" Bernie giggled, "I'll get da gory details from him."

"You do that! I'll bet he stomps your head right up your fanny!" Herol blushed at the colorful language he'd picked up so quickly, but hoped that Black Jack would do just that.

"I kin handle da bum!" Bernie laughed.

"Hi kid, who's your friend?" Tyson asked as they approached.

"Bernie, this is my partner Bob." Herol looked shyly at Miss Savageau.

"Glad to meet you, and this is Miss Savageau, HB's secretary. Marge, this is Herol, my partner."

"I'm happy to meet you, Herol — were you at the scene?"

"Yes'm, he died a little while after we got there."

"I'm sorry," she said, and moved a bit closer to Bob.

"Wisht I coulda seen it," Bernie chirped. "I seen all da big ones get busted up at Roosevelt Field. He stopped abruptly when he noticed Tyson's forbidding countenance.

"Bernie doesn't have a partner — he's a stuntman. I told him that maybe he could practice with us if it's okay with you." Herol regretted what he said, but he'd promised Bernie.

"We'll see." Tyson's voice was far away. "Did Johnnie have anything to say?"

"He was half out of his head, said to tell you hi, and something else, can't remember, but I'll think of it."

"Thought so," Tyson brooded.

"So you're Kid Galahad!" Miss Savageau smiled to brighten the scene.

"That's what Bob says," Herol blushed.

HB had finished his inquiry and made a beeline toward the limousine.

"Everybody down!" someone hollered as the sky filled with a deafening roar. The ground shook and echoed the thunder of a snarling Hisso engine overhead. A flash of crimson wings in the sunlight mirrored the fast departing craft as it reached for the heavens. The handsome Travelair announced "Digger" McGonnigle's arrival.

HB sputtered and brushed the dirt from his immaculate suit. It was embarrassing to be seen on his knees in front of these vagabonds. He gave Miss Savageau a curt nod toward the car and waddled off. It was imperative that he get back to town and make sure the newspaper toned down this story. The accident was bad for business.

"We'll talk tomorrow," HB called back to Tyson.

"That'll be fine," Bob said. He watched them as they got into the car. Miss Savageau looked his way. He smiled slightly and she nodded and smiled back as the car drove away.

The ambulance pulled to a stop in front of the hangar and Black Jack climbed out of the rear. His clothing was splattered with blood. A large blob of gore was smeared across his chest, and red smears covered his hands. There was little doubt as to the grim task he'd completed.

Bernie's eyes took on the dimensions of saucers when he saw Black Jack. He felt light-headed, turned around, and vomited against the hangar wall.

"How about posting a guard by the plane? The goddam scavengers are closing in and fighting for souveniers." Black Jack's voice was barely audible, "I'm going to the morgue and take care of the details. Tell Pop to gas up my Standard. I'm gonna tear the sky loose from this rotten earth when I get back!" He slammed the ambulance door shut and was gone.

Bernie was still gagging as he rested his head against the hangar. Tyson stood nearby, unfeeling, unseeing. Herol realized that he didn't feel sick — he didn't feel anything! Was Tyson's image rubbing off on him? How come Bernie's sick — he's supposed to be a pro!

"Here comes another straggler," someone said.

The sound of the engine was soft at first but it rapidly grew louder. Finally the whole hangar seemed to shake as the plane roared over at rooftop level.

"That's Eddie Gardners Waco," Tyson said. "I guess Eddie's flexing his muscles a bit," he muttered as the blue and silver biplane climbed into the sun.

"Let's get some coffee, Herol. Coming Bernie?"

"Naw, I'll sit this one out," Bernie said weakly. He was still faint from the sight of the blood smeared on Black Jack's clothes.

Herol and Tyson walked slowly toward Carter's office.

"Where'd you pick up that fruitfly?" Tyson asked.

"Met him in the hangar this morning. Claims he's pretty good — heard all about you, too."

"Never heard of him. And he ain't going to practice with us either. He's phoney as a pregnant nun!"

Herol wished he could stop blushing.

Back at the bank, HB drummed his fingers on his desk. He was anxious for Carter to get the show on the road. Johnnie Day's death had been a blow to the morale of the entire group of flyers.

Carter had figured it would take another week to get the group moving but HB wanted them in the air and on their way NOW! He made a decision. The day after tomorrow they would leave — ready or not.

The first stop for HB's flying circus would be Regina, 160 miles southeast. It was an easy two hour flight after which they'd land and spend the rest of the day and evening preparing for the next day's airshow.

Their stay in Regina would be determined by the generosity of the local citizens. If dollars jingled into the company coffers, they'd remain. If not, they'd move on to greener pastures.

From Regina they'd head west to Moose Jaw, Swift Current and Medicine Hat. From there they'd swing south across the border into the United States.

Carter Camp would rule the group with an iron fist. It was imperative that he weld these vagabonds into a disciplined unit. They'd hone their skills and develop the confidence needed to sustain them through the nerve shattering days ahead.

They'd risk their lives from sunup to sunset. They'd walk the wings of their cantankerous airplanes without parachutes. They'd put on mock aerial dogfights to relive their war escapades. And they'd make hundreds of parachute jumps to thrill their audiences.

When they were worn out from stunt flying, they'd hop passengers for whatever the traffic would bear. And finally when they had milked a town dry; and when they were too tired to eat or sleep, Carter would give orders to pack up and move on to the next town.

Eventually they'd work their way across Montana and Idaho hitting Lewistown, Great Falls, Helena, Butte and Missoula. Sometimes they were lucky enough to hit a real grass airfield. More often than not, though, they camped on and flew out of bumpy fairgrounds and rutted pastures.

Their westward trek would lead them eventually to Couer d'Alene, Idaho. There they'd rest and refurbish their planes and their own flagging spirits; but not for long. Rest was a luxury they couldn't afford much of. The northern summers were short, and in order to make as much money as possible, they had to fly as much as humanly possible. Tempers grew equally short under the pressures they faced. The old-timers would lose patience with the newcomers. Arguments would erupt and Carter would have another fight on his hands.

Harry Ross, one of the slickest con artists on the continent, usually preceded Carters flyers to a town by a couple of days. He'd have the local yokels stirred up to a fever pitch by the time the flying circus got to town. Harry's tricks kept him two days ahead of the circus and frequently only two hours ahead of the local sheriff.

HB's flying circus was underway—full throttle ahead!

**An aerial stuntman changes planes the hard way.
(Bill Rhode collection)**

10 HOT TIME IN THE OLD TOWN

The sun climbed against their backs while they flew in in a loose formation westward. The tallest peaks were behind them and lush green valleys beckoned ahead as they followed the tiny thread of water that would lead them to Coeur d'Alene, Idaho, USA. It would be their fifteenth major stopover and a hundred and fifteen airshows without rest. Harry Ross had promised the good folks of Coeur d'Alene that his aerial extravaganza would arrive today and they'd be there even though they were a day behind schedule.

They announced their midmorning arrival with a formation flight down main street at rooftop level—their way of saying hello to the townspeople in the best way they knew how.

Within minutes, cars, wagons, bicycles, and barefoot kids swarmed to the fairgrounds where the weary men circled for a landing. They'd refuel, grab a sandwich if they could find one, and chalk up another half dozen hours before the day ended.

"Who's gonna drive the bus into town tonight?" Tyson asked.

"I think Bernie is, but last time I saw him, he and Pancho was still sellin' tickets," Herol said.

"Maybe somebody else will drive. Sure as heck don't want to get killed on the way into town."

"Ah, Bernie's okay, he's just sorta different. How come he don't ever fly; he's supposed to be stuntman."

"Happy Bottom's too smart to get him killed," Tyson said. "Bernie makes more money selling tickets."

"He told me that Carter was gonna let him jump one of these days."

"Carter ain't that dumb!" Tyson laughed.

Herol decided that he'd better change the subject. "What say we hit up Harry Ross about getting a ride into town?"

"Good idea. Keep smiling and wave at the suckers Galahad — they think we're heroes!"

Happy Bottom sat at his desk. A broad smile covered his pudgy face as he read the weekly memo from Carter Camp. He expected this information on his desk by the end of each week, the financial statement and any problems or otherwise. This latest memo gave the same report as previous weeks — good news! good boys! good airplanes! and good green folding money!

Miss Savageau sat comfortably behind her desk, a faint smile lingering on her pretty face as she read the homely scrawl of Tyson's letter:

Dear Marge

I don't expect you to answer because it's hard since we're not in one place long enough. I think of you a lot. Me and Herol are doing good. We're top billing wherever we go. The crowds go crazy over the kid, and he believes all that hero stuff, and I have to knock him down a peg or two every few days. We're still heading west but will be backtracking soon. We're in a town in Idaho, USA. I can't spell the name, maybe it will be printed on the envelope.

We all need a rest, and the planes too. I don't go no place, but fly and sleep. Nobody got killed yet, but we all got lots of bruises and we got lots of iodine and rags around to keep patched up. I think of you every day and wish I could see you again soon. Maybe you would like to have dinner with me the day we get back, I guess that won't be for a couple months at least. I promise to have you home by midnight like you ask. You're the nicest girl I ever knowed.

Your friend
Bob

She tucked the letter carefully into her purse. "That sweet man, he's so shy and so quiet. Maybe that's what frightens me. Come home soon, Bob," she whispered.

The Stockman was a typical small western hotel, a two story wooden structure with an open porch, littered with ancient wooden chairs for the guests to sleep in, gossip, or idle their time away. The only attempt to modernize the place was the addition of a false front made of imitation brick. An electric sign straddled the sidewalk and flashed, "Gent's Sleeping Rooms — Eats." Twenty single rooms lined the upstairs hall and five on the ground floor. The remainder of the building consisted of a lobby cluttered with horsehair sofas and spitoons, and wrapped in faded wallpaper. A side door led into a dining room, a large room with an oak counter near the wall and tables and chairs strewn at random around the room. A large round table occupied the center and was usually used for socializing.

The place was owned by a robust couple, Mr. and Mrs. Gietzentanner. Everybody called the old man "Geetz" and his wife, "Mrs. Geetz." They'd never bothered to shorten their name, so the townspeople did it for them.

Mrs. Geetz served tasty boardinghouse meals to the cowboys that came into town on business or a bit of hell raising. Some of the airmen also called this home during their stay, while others found places elsewhere more suitable to their individual tastes. A few diehards slept under the wings of their beloved craft as they always had. The more fortunate, like Carter Camp, enjoyed the hospitality of the sheriff's home. Others like Harry Ross drifted into the hotel occasionally at meal time, then faded into the night.

Tyson, Herol, Bernie, Gimpy, and Russ called the Stockman home.

Mrs. Geetz loved to cook and was delighted when a guest asked for seconds. She'd scurry back to the kitchen and return with a larger portion of the day's special. Mr. Geetz allowed that his wife's cooking was the best in town, but much to his regret she was also the most generous. Sometimes he broke even, depending on the size of Mrs. Geetz's meals. She kept an ample reserve from the evening meal for the aviators who drifted in between 9:00 and 10:00. They'd lived on coffee and cigarettes since dawn and would be famished. She looked forward to their arrival. They'd surely be hungry enough for "seconds." Poor Mr. Geetz agonized to see his profits go down the drain, but he'd smile and say nothing.

Gimpy was the first to show up. He'd lucked out and caught a ride with the sheriff and Carter. He went to his room long enough to wash the day's grime off his face, slick his hair down, and beat the excess dust from his jacket and breeches. He strolled through the restaurant to the far wall and slumped in a chair by a small table. He lit a cigarette and relaxed for the first time since sunrise.

"You reckon he's one of them dudes, or maybe a city slicker?" the cowboy at the card table asked.

"I dunno," his companion replied. "He's too dusty lookin'. Besides, his duds ain't fancy like them city fellers wear."

Gimpy overheard their conversation and turned his head in their direction and gave them a friendly nod.

The first cowboy returned his nod and the second raised his hand in a friendly gesture.

"He's right friendly; I reckon he's all right."

"Good evening, Mr. Gaines! You want to eat right away, or are you waiting for some of the other flyers?" said Mr. Geetz.

"He's one of them flyer fellers — they been doin' tricks at the fairgrounds all week," one cowboy said.

"Somethin' is writ on his back," the other cowboy drawled.

"It says Gimpy somethin' — can't make it out."

"The coffee sounds good for now, I'll wait, thanks," Gimpy replied.

Upstairs, Herol and Tyson were getting ready to go down and eat.

"Ain't you got any cleaner lookin' duds than that?" Tyson asked.

"Why sure! But I'm hungry!" Herol groaned.

"Well, put 'em on! We're top dogs in this flea circus and we gotta look the part. Put on a clean white jacket and breeches, shirt and tie too — and dust off them boots. I don't give a dang if you stink underneath, just so you look good from a distance!"

"Okay," Herol grumbled.

Tyson gave him a rare smile. "See you downstairs."

As Tyson walked into the dining room, the cowboys took notice.

"Here comes another of them flyer fellers!" The cowboy nudged his buddy.

"He's sure a ornery lookin' critter! Lookit them eyes!"

Tyson glanced around the nearly deserted room, spotted the cowboys staring at him, and muttered a vacant "hello" as he passed by their table.

"Don't that beat all! It says 'The Great Diablo' on his back – name sure fits him too!" the first cowboy whispered.

"Hey Bob! Care to join me?" Gimpy waved him toward the table.

"Sure, thanks," Tyson said. "It's been a long time between breakfast and supper."

Mrs. Geetz hurried from the kitchen. "I brought coffee for both of you."

Bernie eased the bus around the corner of the hotel, then ground to a stop in the narrow alley where he'd park the old clunker for the night. He'd let Harry Ross, his only passenger, off a block away. Harry disappeared as always to join the night people. Bernie felt good tonight even though he was worn out, dirty, and hungry. It was a wonderful day! He and Pancho had a bet going — only two bits — who would sell the most tickets today. Pancho did a land-office business, but he was pretty sure that he'd won the bet. He'd sold over seven hundred! Since he hadn't made the grade as a stuntman, he'd proven that he could still make money for the circus. The friendly pats on the back told him that he was with friends. He didn't feel left out anymore. The delicious meal Mrs. Geetz had waiting for him made his mouth water. He was too tired to get cleaned up. He locked the door of the bus and made a beeline for the front door of the hotel.

Herol doused his head with water from the basin, wiped the excess off with a handtowel, and draped it over his shoulders. He crouched in front of the mirror and pulled a comb through his dark hair, then pasted the unruly stands down with his hands. He dug into his ears with the clean towel and rubbed some of the day's grime from the back of his neck. The once fresh towel resembled a floor mop, and with a guilty feeling he folded the blackened side from view and hung it back on the wall.

"I'm so damned hungry, I could chew the leg off an angry bear!" He slammed the door behind him and sprinted down the stairs.

"Hey Galahad!" Bernie shouted as he pushed through the front door.

"Hi Bernie! I saw you out there herding all those cowpokes and their girlfriends through the gate — you sure had 'em corraled!"

"Yeah, it was a good day. I saw you too, up there climbin' down that rope onto Black Jack's wing. Scared heck outa me!"

"Me too!"

"How come you look so sharp?" Bernie looked at his own dusty apparel.

"Tyson makes me slicker up all the time. I'd have eaten supper by now if it wasn't for him."

"He's right, youse is da big time, I'm just a flunky but Carter sorta promised . . ." His eyes, half concealed behind dusty glasses seemed to cloud over a bit.

"You'll get your chance. You're making so much money for Happy Bottom that you'll be vice president soon. How about supper with Bob and me?"

"Ah dat Bob don't like me so pretty good. I gotta clean up some, you look so fancy and I look like a bum."

"Ah, these cowboys smell as bad as their horses, let's get something to eat."

Bernie made a futile attempt to shake some of the dust from his sweat stained shirt and Herol slicked back his hair once more as they headed for the dining room.

"Well I'll be danged! Here's a couple more them flyer fellers!" The cowboys were getting their share of free entertainment tonight.

"That fancy feller in them white duds don't look like he does nuttin' — that other feller's been worked to death!"

The 'flyer fellers' stood at the entrance for a moment before walking in. Gimpy and Tyson were wrapped in conversation while they devoured huge quantities of food. The cowboys sat behind their neglected cards and stared at the newcomers. Russ Sorensen was getting up from the counter as he took a last swig of coffee.

"Hi Russ," Herol and Bernie chorused.

"Hi guys! Might be back later for a nightcap." Russ gave them a nod and left.

Gimpy and Tyson looked up from their meal, waved, then resumed their conversation.

"Looks like they're talkin' business. Let's grab that table behind the cowboys." Herol said.

"Hiya!" Bernie greeted the cowboys as they edged past their table.

"Hi!" they replied.

"Kid Galahad — it says on his back — sure fits! Just like that critter in the black t'other table."

"That other feller ain't got no name — sez 'OFFICIAL' — he must be one of the top hands. Might be they'd play cards later on?"

"Reckon they might give us a airyplane ride? If we won a few hands of poker, it might be worth it."

The Pink Pussycat nestled in a growth of trees about five miles outside of town, an unpainted cabin that advertised "Dancing". A dark clearing in the back served as a parking lot as well as an arena where the befuddled customers held their nightly brawls. The interior was a large smoke-filled room, crowded with tables and a half dozen booths against a wall. A battered player piano and a gaudy nickelodeon crowded the tiny dance floor near the rear exit. Dim shafts of light glowed faintly from a ceiling fixture through clouds of cigarette smoke and filtered between the shadowy figures at the tables. Every table was cluttered with pop bottles, glasses, and ash trays piled to overflowing. Under the tables, bottles of moonshine were tucked safely between the boots of the cowboys.

Two nervous waitresses squirmed between the customers, dodging the amorous fanny pats. A big bruiser stood near the front entrance, arms folded across his chest and itching to throw some luckless drunk through the back door.

Black Jack and Dutch Hauptman sat in a corner booth near the dance floor. Neither spoke, nor showed much inclination to. They drank their booze in silence, each immersed in his own world. Black Jack was still haunted by the loss of his friend, Johnnie Day, and needed moral support but couldn't find it here, or any place else. Dutch was getting too old for this kind of life and longed for that never-never day of a peaceful old age.

"What say we go to the Stockman, some of the gang will be there," Black Jack mumbled.

"Ain't we gonna get drunk? My tail has been draggin' for weeks—time we did some hell raising," Dutch said. "Sure feel like a good time tonight, if I could get somebody to fly for me tomorrow. Maybe Russ will." Black Jack didn't pay any attention to Dutch, but waved to the bouncer.

"Whaddya want?" the bouncer growled.

"Wonder if you'd call a cab for us to get back to town. We ain't got no transportation," Black Jack asked.

"What the heck do I look like, your butler? Just a minute, be right back. There's a couple bums I gotta throw out." The bouncer cut a swath through the tables till he reached the far side of the room. He pulled two noisy drunks off their chairs, grabbed the back of their pants, and carried them out like two chunks of beef. The rear screen door clattered, then there was a thud as the drunks sprawled in the dirt.

The bouncer bulled his way back to the table.

"Like I was saying, I ain't no damned butler." The bouncer breathed hard to impress his audience.

"No, you ain't my butler but you git goin'! I want a taxi!" Black Jack stood up and glowered.

"Better do like the gent says or he'll twist your scalp off and hand it to you," Dutch said.

"Awright," said the bouncer. "But I better not see you clowns in here again."

"Git goin'!" Black Jack growled and sat down, mindful of the furtive glances from nearby tables.

Dutch upended the pint of moonshine into their glasses and set the bottle on the table. They gulped their potent drinks in silence and glowered at the crowd.

"Awright you guys, the taxi is here. Get that bottle off the table. Booze is supposed to be on the floor, only setups on the table."

"Ah knock it off! The bottle's empty," Dutch countered.

"Get out before I throw you out!" the bouncer shouted for all to hear.

A hush fell over the crowded room. Curious heads turned in their direction at the promise of more excitement.

Black Jack hadn't acquired his name by chance. Tonight was just one more time he'd justify his ominous title. He swung a wicked right into the bouncer's gut that doubled him over to the floor, then grabbed the back of his collar and the seat of his pants and gave a powerful heave. Two hundred pounds of muscle and fat sailed across the dance floor and banged through the screened exit.

Black Jack lumbered across the floor, acknowledging the cheers and delighted yells that rose through the room. Dutch waved to the audience and slammed the screen door behind them. Moments later, the door banged open and the bouncer sailed across the dance floor again and wound up in a heap under the player piano. Black Jack walked back in and over to the piano. Dutch followed him dragging the two drunks who'd just been thrown out.

"What did you do that for?" the bouncer hollered as his opponent towered over him.

"Wanted you to see how your job looked from the other end," Black Jack growled.

"You're a lousy bouncer! Better do that last job again." Dutch said as he piled the two drunks from the parking lot on top of him.

They gave a brief nod to the crowd, turned their backs, and headed for the waiting cab.

Russ closed the door of his room and descended the stairs to the lobby. The long hours in the cockpit had made him too restless to sleep. There might be other restless ones downstairs.

"Evenin' Mr. Sorensen," Mr. Geetz nodded as he walked in.

"Evening — Hi Dutch" Russ greeted the big mechanic as he stumbled through the door.

"Hiya!" Dutch hung onto the door and swung himself around to peer into the alley after his buddy.

The tranquil atmosphere that prevailed was shattered by a muffled crash midst the scuffling of boots and angry shouts from the restaurant. Mr. Geetz rolled his eyes heavenward as he left his desk and hurried toward the commotion. A tense scene was on the verge of erupting. Herol, Bernie, and the cowboys had gotten together for a game of poker and had devised a unique system of gambling. The color of the poker chips represented five, ten, or fifteen minutes rides — horse or airplane, depending on who held the chips.

For a time they got along fine. The cowboys spun yarns about rustlers and roundups while the airmen reciprocated with tales of air circus derring-do. The cowboys, accustomed to long hours at the card table, could lean back on their chairs and, without effort, balance on the hind legs as they played. Not to be outdone, the airmen attempted to show their expertise also. Winnings were fairly well distributed, except that the blue chips were cornered by one of the cowboys. He must have had half a day's flying time stacked in front of him. Bernie had just told an improbable story, and since it amused him as much as his listeners, he giggled and reared back on his chair, lost his balance, and crashed to the floor. Herol followed suit when he caught his knee under the table and pulled it over, following Bernie to the floor. Poker chips, cowboys, aviators, and furniture were scrambled together in a heap!

"You did that on purpose!" the winning cowboy roared. He didn't know who to blame, so he glared at them both.

"Ah, it was an accident," Bernie argued.

Bernie's glasses took a flight of their own when he made his rapid decent from table to floor, and as he reached for them, the angry cowboy mistook his intention and threw a haymaker at Bernie's jaw.

A free-for-all erupted like a case of dynamite. Gimpy and Tyson leaped into the melee to break up the fight. A cowboy took a wild swing at Tyson and missed, his fist landing square on the side of Gimpy's head sending him crashing against the wall.

With Bernie disposed of for the time being, the first cowboy upended the table high into the air and sent it crashing on top of Herol's back. "Kid Galahad" felt like a mashed bug sprawled under the table with the cowboy standing on top of him.

Tyson was trading punches with his cowboy when Mr. Geetz, Russ, Dutch, and Black Jack rushed through the door.

"Hot damn! Here we go again!" Black Jack roared.

"Let's break it up, you nitwits," Dutch hollered.

"Oh my, oh my," was all that Mr. Geetz could say as he circled warily around them and disappeared into the kitchen.

Russ made a flying tackle at the cowboy standing on the upturned table and sent both of them sprawling into a corner.

The old cattlebuyer sitting in the corner by himself grabbed his coffee pot, stood up, and bounced it on Russ's head. He returned to his chair and continued to watch the show.

Herol struggled to squirm out from under the table, only to change his mind when a leather boot stomped on his knuckles.

Gimpy shook the cobwebs out of his head and made an effort to get up but his wooden leg creaked in protest and collapsed under him. He pulled his boot off and examined the splintered leg under his breeches. It was like holding a handful of toothpicks. If he couldn't stand on it, there were other uses for it. He watched Tyson put up a gallant but losing battle against his younger opponent, then redoubled his efforts to unstrap the wooden leg from his padded stump.

Dutch rescued Bernie's glasses, then grabbed a pitcher of water from the counter and poured it on the listless figure sprawled in the middle of the floor. He threw him over his shoulder and carried him into the lobby and dumped him into a chair.

Tyson couldn't hold out much longer against his cowboy and slowly backtracked toward Gimpy. The time was right. Gimpy crawled across the floor, hoisted himself to a standing position and swung the wooden leg like a ball bat. The cowboy slumped into Tyson's arms and they both melted to the floor.

Black Jack stomped across the table and mashed Herol into the floor again as he bent over to part the embattled Russ and his cowboy. They both swung at him without looking, so he grabbed them by the hair and bounced their heads together. While birdies chirped inside their aching skulls, he picked up their limp bodies by the seat of their pants and stomped once more across the table and the hapless Herol.

"Where you want I should dump these guys?" he hollered over his shoulder.

"In the lobby, I'll get them to their rooms later," Mr. Geetz shouted from the kitchen.

"Line him up with the others!" Mr. Geetz shouted again as Dutch carried an unconscious cowboy through the door.

The show seemed to be about over, so the old cattlebuyer rose from his table to go to bed. On his way out he passed the upturned table and the still imprisoned Herol, struggling to free himself.

"Galahad?"

"Yeah."

"You flyers is all crazy," he said and stomped his boot on Herol's knuckles.

"Hey! Cut it out! Help me get out from here."

"You're pretty good at tricks, this one ought to be a cinch!"

Black Jack strode over toward Gimpy and Tyson who were struggling to their feet, while Dutch intercepted the cattlebuyer at the door.

"You shouldn't step on a man when he's down!" Dutch growled as he kicked a well-aimed boot to the cattleman's backside.

"Ooooooooo—!" the cattleman bellored as he shot through the door, his big hands spread protectively across his bottom.

Ten bedraggled characters were piled in a semi-circle of chairs and sofas. Each one stole a sheepish look at his neighbor as they compared battle scars. A few made self-conscious efforts to grin through swollen lips and battered eyes.

Mrs. Geetz emerged from the kitchen clucking like a mother hen. She carried a pan of water, soap, and towels for her boys.

Mr. Geetz nodded attentively as he listened on the phone to a barrage of questions from the sheriff. "We don't have any bouncer, sheriff. I'm the boss here — and Mrs. Geetz. Just a minute, I'll ask around. Anybody here throw a bouncer out tonight?"

A burst of laughter erupted.

"We threw everybody out except the bouncer!" Bernie chirped through swollen lips.

"Tell the sheriff to send over the bouncer. We'll throw him out if nobody else will!" a cowboy shouted.

"No we don't have no trouble here, sheriff. We're all good friends — okay — sure goodbye."

"Everybody here come out to the fairgrounds tomorrow for a free airplane ride on me," Black Jack roared.

"Yippee!" yelled the cowboys.

"And if we don't get killed in the airyplane, how about everybody comin' out to the ranch Saturday. We'll have a big barbecue and horseback ridin' too."

A chorus of approval filled the lobby.

The old cattlebuyer got up gingerly, still holding a hand to his backside. "How about everybody shakin' hands? It was a good fight and we ought to be friends now."

Dutch made the first move, he came over with his hand extended. "Shake partner!"

Mrs. Geetz interrupted the handshakes occasionally as she tended a swollen jaw or black eye.

Tyson and his former opponent were talking when the cowboy reached inside his pocket and threw a ten dollar bill on the floor. "How about everybody match that, just in case we did some damage during our card game."

Bruised knuckles dug into pockets and scattered bills in a heap on the floor.

"Anybody got an idea how we can fix Gimpy up with a new leg? He sure saved me from getting clobbered, though my buddy here has a heck of a lump on his head." Tyson grinned at the cowboy.

"Why shore! We'll take Gimpy with us to the ranch tomorrow. We'll make him a new leg, better than the old one. We got a blacksmith shop in case he needs new parts, and lots of good wood."

"What say Gimpy?" Tyson asked.

"Heck! Even the termites wouldn't want the old one!" Gimpy complained. The front door opened, and the sheriff and Carter walked into the lobby.

They stared in disbelief at the array of battered faces that squinted back at them.

"What happened? You guys come out second best with a buzzsaw? Thought you said you didn't have any trouble here," the sheriff glared at Mr. Geetz.

"We're all friends — see?" Mr. Geetz pointed in the direction of several battered heads that wagged in approval.

"Are you sure nobody here threw out a bouncer?" the sheriff asked.

"I threw a guy out at the Pink Pussycat. The bum deserved it anyway," Black Jack admitted.

"I told you, we ain't got no bouncer here," Mr. Geetz interrupted.

The sheriff made an exaggerated gesture of understanding, then asked, "How come you guys are so beat up? — and Mrs. Geetz, how come you're patching 'em all up?"

"They're always playing, and boys always get rougher than girls when they play." Mrs. Geetz put on her sweetest smile.

"Sure! Boys play rougher than girls," the sheriff mimicked.

Carter, mindful of tomorrow's show, looked over his flying menagerie. Secretly, he wished that he'd been in on the excitement. A faint smile passed his lips, then quickly vanished.

"What happened to you, Bernie?" Carter asked. "Run into a doorknob?"

"Nah, nuttin like dat, was playin' cards and fell down — musta banged myself up. Lost my glasses and couldn't see nuttin'," Bernie said.

Carter thought about the weekly report that Happy Bottom would be looking for. He'd give him the same "good boys, good airplanes, good money." Happy Bottom wouldn't be interested to hear that the boys played cards and Bernie lost his glasses!

"Next time don't fall down so hard!"

"What's all that folding money on the floor — gambling?" The sheriff pointed an accusing finger toward the floor.

That's for Mr. and Mrs. Geetz, sheriff. We owe them some money—you know—rent, food, wear and tear," said Tyson.

"Yeah — you guys sure look like you had some wear and tear," the sheriff grumbled.

A chorus of "Yeah! Sure! You bet!"

"I must have got that story about the bouncer all wrong. He didn't get throwed out, he musta got throwed in! Besides, the Pink Pussycat is in the next county." The sheriff winked at Black Jack.

"Well, take it easy fellers, and don't play so rough. Somebody might get hurt." The sheriff opened the door for Carter and they stepped into the night.

"GOOD NIGHT SHERIFF!" they all roared.

When the stuntman's day of wing walking and parachuting was over, the grand finale was often a car to plane change. (Bill Rhode collection.)

11 FIRST SOLO!

"Watch where you're goin', you nutty squirrel!" Tyson roared at Bernie, who wrestled the steering wheel of the bus as they bounced along the gravel road.

Bernie let up on the accelerator and they slowed down to a gallop. "How's dat?" He turned in his seat.

"That's more like it! Watch the danged road!"

"Yeah — only a couple more miles to the hotel."

Herol held his arms tight against his leather jacket and groaned to himself, hoping that Tyson wouldn't notice.

"What's the matter kid, you don't feel good?" Tyson's wrath had tempered to gentle concern.

"I think I busted some more ribs on that third jump. I got banged up when I bounced off the grandstand roof. It looked like I was gonna' land in the crowd, and I spilled so much air out of the chute it nearly collapsed. I musta passed out because I don't remember hitting the ground."

Tyson eyed his partner with mixed emotions of concern and anger. "Why didn't you call off that last jump? The suckers had enough for one day anyhow."

Herol nodded in agreement and winced again as Bernie plowed across a corrugated stretch of road.

"You owl-eyed squirrel! You and me is gonna' have a butt kickin' contest when we get off this bus and I bet I win!" Tyson pounded the seat in front of him with rage.

Bernie didn't answer and remembered not to turn around. Instead, he sat rigid and watched the road ahead. He hoped fervently that the contest would not be held.

"Those the same ribs you cracked a few weeks ago?" Tyson asked.

"Nah! They're better. Now it's the right side that's busted."

"Sure wish you hadn't made that last jump, but we're gonna get a few days rest. Carter says we're gonna head further west and Harry Ross has already left to get us set up for the next stop."

"Where we goin'?"

"Beats heck out of me. I hear that we're gonna' keep goin' till the water runs in the top of our boots. Then we know we hit the Pacific Ocean." Tyson gave one of his rare smiles that made both of them feel better.

"I got so many lumps, I hurt all over, and the black and blue spots switch places every day. How about that rest you mentioned?" Herol smiled and wished he had tape on his ribs.

"Not countin' today, we get three days to goof off. We ought to spruce up the Eaglerock a bit and look over our chutes and the rest of the gear, but we should have plenty of time for hell raisin' or sack time. How's that grab you?" Tyson smiled.

"Ain't no flies on that program," Herol agreed.

"We'll get those ribs taped up soon as we get to the hotel, provided that goofball doesn't kill us in the meantime." Tyson glanced in the direction of Bernie.

Bernie announced their arrival with a clashing of gears as he tried to coax the gearshift into neutral. His surefire remedy of the past always worked, so he turned off the ignition and the wheels jolted them to a stop. He sat rigid as the expected barrage from the rear of the bus exploded around him.

Tyson waved his arms in disgust at the world in general, then stomped out of the bus.

"Meet you for coffee in about half an hour, okay?" Herol asked Bernie as he stepped out.

"Yeah, okay," Bernie agreed.

Herol quickened his step to catch up with Tyson who was turning the corner at the end of the alley that led to the hotel entrance.

Bernie followed at a safe distance lest he encounter Tyson and incure his wrath again.

Herol slouched comfortably in the easy chair, his dusty boots propped on the foot of the bed. He ran his fingers over the wide strips of tape that encircled his rib cage and thought of the promised three days vacation.

Tyson was lost in thought as he sat on his bed. He looked up from the sheaf of papers in his hands and asked, "How the ribs feel, kid?"

"They're okay. I can breathe good with the tape on them. Sure glad we're gonna get a rest."

"Get them danged boots off the bedspread," Tyson muttered softly.

"Ah Bob, you used to do a lot worse when you was hittin' the bottle. I guess I'm tryin' to be like you so you have a real partner for a change."

Tyson sat upright and his gentle mood vanished. "You wanna' be like me! You wanna be a bum?"

"Ah, you ain't no bum. You're the best around. Everybody says so."

"Yeah, the best! No home, no family or friends," Tyson said softly to himself.

"We're friends, ain't we?" asked Herol.

"Yeah we are, but you don't understand. What do you want out of life?"

"I wanna be the best, like you."

"You see, kid, you're dumb like I always tell you." Tyson's words hardly matched the soft manner in which he spoke.

"Yeah, guess I am."

"You miss the point, kid. You ain't dumb in the air. You're gonna be better than me someday. You're dumb on the ground. You don't know nothin' about life."

Herol had never seen the gentle nature of this complex man assert itself as today. Why was he so harsh and unfeeling most of the time?

"Bet you ain't ever chased a broad — uh — went on a date with a girl. Ever had a girlfriend?"

"Yeah, I had a girl in Fargo, but she always got mad at me 'cause I always did the wrong thing."

"What did you do?" Tyson prodded.

"Ah, just dumb things. I don't understand girls."

"What the heck! She's just another broad," Tyson laughed.

"Bet you don't call Miss Savageau a broad. She's too nice to be a broad."

"That's right! She's a lady and she ought to go out with a gentleman instead of a bum like me. What's with this girl you had?"

"Ah nuts! I took her to a movie and we sat in the balcony where all the guys and their dates were. There was some real mushy love movie playing and all the guys was smoochin' with their girls. My girl was lookin' at me kind of cross-eyed 'cause I never did no smoochin'. I didn't know what to do so I grabbed her quick to give her a smooch and knocked her hat off. She got sore when she couldn't find her hat and the other girls was startin' to laugh at her, then she hauled off and poked me."

"You're sure a great lover," Tyson chuckled. "Did you get things patched up with her?"

"Sort of, but she got mad at me again. I saw her downtown and I asked her if she would like to see our flying field, and I told her about the planes too. She thought that was great, so I took her out there on my motorcycle and that's when she got mad again. I drove too fast and her dress blew up. I saw her pants! The gravel on the road messed up her fancy shoes and silk stockings. I knew she was mad so I figured I best get busy. I told her I had to clean one of the planes — the Travelair that Dick got creamed in — and I'd wash if she'd wipe."

Tyson shook his head in disbelief, then covered his face with his hands. "Yeah, you wash and she wipes — hoo boy!"

"Well, she got her dress messed up from wiping, and I told her I'd wash it when I did my coveralls at the end of the week. Then she really got mad! So I took her back to town and when she got home I thought she was gonna take a poke at me again. I can't figure out girls!"

"You sure as heck can't! You know too much about dying and nothing about living. We're gonna have to get you educated." Tyson laughed for the first time in months.

"What we gonna do this afternoon? Don't seem right that we got all this time off," Herol asked.

"Thought we'd take care of a little business, then we can goof off for a couple days. I been keepin' track of the time on the plane and I gotta bring the log books up to date. Then I'm gonna make a guess on the gate receipts and our expenses and come up with an estimate on how much dough we made so far. I bet we clear about ten grand each so far and the season ain't half over yet."

"Wow! Let's buy a couple planes and go barnstorming on our own next year!"

Tyson's reply was meaningless and addressed to no one in particular. "How many sunsets left?"

A sense of dread engulfed Herol as he recalled Johnnie Day's forgotten words.

"Yeah, Johnnie said something like that the day he got creamed. He said I should tell you but I forgot. Guess I was too nervous to remember."

"Yeah, I know. Forget what he said, okay? Would you like me to do the parachute jumping till those ribs heal. You got a bad ankle too, ain't you?"

"The ankle's okay. I keep it wrapped tight and when I

pull my boot on, it gets so tight that I don't feel nothin'."

"That's what I thought," he smiled and continued in a gentle tone. "I brought your parachute log up to date and counting today's three drops, you got 187 jumps under your belt. You gotta be pretty good to press your luck that hard. If you keep it up, you'll have over 300 by the end of the season — provided you ain't crippled or dead. I'll trade off with you for a while and do the jumps, and you can do all the flying."

"That's great! I don't have to wear that chute, and them harness burns on my shoulders and legs will heal up too!"

Tyson rose to his feet and approached the youth. He placed his hand on Herol's shoulder and gave him a friendly nudge. "You're okay. I'm proud of you."

"Gee thanks! You never said that to me before. You never said anything nice before!"

"Aw shut up! You keepin' your pilot log book up to date?"

"I ain't got one. Thought you was doin' that. You said you was gonna do all the thinking. Guess I shoulda asked about that."

"You got me there. Guess I shoulda done it for you. You got a log book from when you was in Fargo?"

"Gee, I don't know! The log books in the Travelair and the Standard will have something."

"Maybe for now we can go back through the Eaglerock's log and figure out some pilot hours and this fall we can stop in Fargo and get you credit for that time. Remember the date you soloed in Fargo?"

"Ah, I don't know no dates. I don't know if I soloed or not," Herol replied.

Tyson's mood changed abruptly. "What do you mean?" Everybody knows when he soloed! That's the biggest event in any pilot's life!"

Herol searched for words, fearful of Tyson's wrath. He knew what he wanted to say but couldn't find the right words.

"Me and Pete Marcus and Howard Lerom was always flying. We'd go up together to share the cost of the gas. Sometimes I flew and sometimes they did."

"Did you fly without dual controls?" Tyson asked.

"Sure, but we never kept track of stuff like that."

"How may danged hours you been in the air and nothing to show for it?" Tyson stormed around the room, then sat down on the bed.

"We didn't need none of that stuff until today, how come?" Herol ventured.

"Because we're a bunch of outlaws and if you ever expect to keep flying, you got to get legal. The Provincial government is starting to raise hell and the Feds in the States are too."

"Oh yeah, I remember one time for sure when I soloed — or maybe I didn't, 'cause I was carryin' a passenger. Kiddie Karr was there when it happened."

"Nuts!" was all Tyson could say. He waved his hand for Herol to continue.

"One day, Dick came out to the field and asked me to go with him to Grand Forks. He had a case of home brew to deliver to a friend of his. Well, I had to take the dual controls out of the front cockpit to make room for me and the beer. Soon as we was in the air, Dick hollered for a bottle of beer 'cause it was such a hot day. We was about ten feet off the ground chasin' coyotes across the prairie. I got kinda scared when he rolled the wheels in the grass and jumped over those critters. Dick threw his bottle at one of them and hollered for more."

"Hey! You nuts, or am I? What about the solo flying?" Tyson groaned with despair.

"I'm comin' to that. I gave Dick another beer and I took one too 'cause I didn't think we'd make it to Grand Forks. I don't know what happened, but I peed in my pants. I don't know if it was the beer or if I was scared. What do you think?"

"I don't give a dang if you peed in your hat! Get to the point." Bob groaned again.

"Well, by the time we got to Grand Forks, Dick was so drunk we hit the wire fence on the north end of the field and tore the fabric on the lower wing."

Tyson interrupted, "How about the solo, kid?"

"Well, I had to use my shirt to patch the wing and I stuck it on with tape. Dick sat in the front cockpit and finished the beer. He passed out so I cranked up the Travelair and flew us home. Kiddie Karr met us when we got home and Howard Lerom was there too."

"What a solo! If the Feds ever got wind of that, you and Dick would be buried under the jail!"

"I guess that was a sort of solo, wasn't it?"

"Don't ask me! Maybe the Feds would know, but don't ask them!" A faint smile was hidden behind Tyson's hands.

"You get back to the field this afternoon and fly solo for a while and I'll get a log book figured out for you somehow. How's those ribs?"

"Ah, they're okay I guess."

"Good! Get your buddy the squirrel to drive you to the fairgrounds and he can help you get cranked up. Put a sack of bricks, the tool box, or something that will weigh about as much as two passengers in the front cockpit. Then go fly your tail off all afternoon."

"Gee, I forgot, I told Bernie we'd have coffee. Okay, I'll go fly awhile. See you for supper."

After they got to the airport Herol pumped gas from a barrel into the Eaglerock's tank while Bernie wrestled the heavy tool box into the front cockpit.

When Bernie got the tool box in, he kneeled on the wing and looked at the instruments and controls in the plane. He wiggled the stick back and forth.

"Cheez — I'd sure like to go wid you. I don't get no chance to fly. What say?" Bernie pleaded.

Herol made an effort to appear busy while he thought how to reply. He hadn't liked Bernie when they first met, but they'd become good friends and buddies over the past several months. Why not! It's only going to be a solo anyway, he told himself.

"Sure thing! Maybe we'll scare up some excitement. You wanna wear my chute? I got some sore ribs, and besides, there ain't no cushion in front. You should have plenty of room beside the tool box."

"How about dat! I'm ready!"

Herol stepped down from the wing and coiled the gasoline hose around the barrel and hauled it away to the sidelines. He leaped onto the wingwalk and with another step he was in the rear cockpit.

"Give 'er a crank Bernie — switch is on."

Moments later, the Eaglerock trailed a plume of dust as they raced across the grass. The plane lifted gracefully and off they flew.

Herol squirmed with excitement over his unexpected luck. He wondered idly what they could do for excitement, then closed the throttle and shouted against the wind.

"What you wanna do, Bernie?"

"I'm wid you. Just so it's fun!"

"I got an idea. Wanna jump?"

"Who me? Gee, I dunno."

"Hey! You're wearing my chute, and I fresh packed it this afternoon."

Herol laughed as he advanced the throttle to regain altitude. The jump idea sounded pretty good. This 'first' solo would be some fun after all.

He tapped Bernie on the shoulder and with a questioning look, pointed to the ground. Bernie gulped, reached for the center section struts and pulled himself halfway out of the cockpit. He stood rigid in the face of the propeller blast, too scared to move. His big moment had arrived and he couldn't live up to his expectations. He relaxed his grip on the struts and sank into the safety of the cockpit.

The Hisso slowed down to a soft rumble as Herol retarded the throttle and shouted.

"Hey Bernie. Wanna jump like I do sometimes — from the top of a loop? It's easy!"

"I dunno, maybe," Bernie shouted.

"It's a cinch! Tighten your harness good, especially the leg straps, and unbuckle your seat belt."

"Yeah, okay! I ain't got no belt on. It's under the tool box."

"Okay, relax. I'm gonna make a loop. I'll flatten out on top and your fanny will start to leave the seat. Then you'll be on your way."

"Yeah, okay, I guess."

The Hisso growled as they picked up speed. He aligned the nose on a gravel road below them and listened for the wires to sing their exact pitch. He eased back on the elevators and watched the nose climb through the horizon. The blue sky enveloped them as the arc of the loop sped them straight for the heavens. Herol kept a steady back pressure on the elevators to keep Bernie glued tight in the seat as they reached for the clouds. He bent his head back to catch sight of the horizon and earth below as the Eaglerock raced past vertical and continued its inverted arc. A split second away they'd reach the apex of the loop and flatten out. He was confident Bernie's first jump wouldn't be like the one that Bob surprised him with.

"HEY, WHAT'S HAPPENIN?" Bernie yelled as the tool box leapt off the seat and disappeared through the fabric of the center section wing above his head. A split second later, his head was imbedded in a maze of splintered ribs and torn fabric. The tool box and its contents had long disappeared and was floating earthward. Bernie kicked and yelled as he tried to get out of his perdicament. He appeared to be doing a handstand, then slowly his head emerged from the inner recess of the wing and he tumbled backwards and out of the airplane.

Herol half rolled from his inverted position and pulled the Eaglerock into a steep spiral to follow Bernie's descent. A white streamer appeared as the canopy mushroomed above the tumbling figure. He breathed a sigh of relief as his buddy descended toward the center of the grandstand area.

"I'll sure get my tail twisted when Bob hears about this," Herol said as he continued to spiral around his sky-jumping buddy. "I forgot all about that front cockpit being under the center section. Gee! When Bob dumped me out, I was in the back! Those tools will be scattered all over, too!"

"YYYEEEOOOWWWWW!" Bernie yelled at the top of his lungs.

After the heart-stopping terror he'd experienced on falling out of the plane and after the painful shock of the chute opening, Bernie was feeling incredible. The sheer beauty of floating above the earth was breath-taking. Down and down he drifted.

Herol touched down with the big biplane just as Bernie hit the ground like a sack of bricks.

Bernie was lying flat on his back exactly where he'd hit when Herol taxied up. Herol switched the magnetos off and jumped out of the cockpit as the plane rolled to a stop.

"BERNIE!!" he yelled as he ran up to his partner. "BERNIE! YOU OKAY?"

Bernie opened his eyes. "Dat was wondaful! Dat was beautiful!" I'm okay. I'm just tinking how great it was."

"Wow!" said Herol. "I though maybe you were hurt. If you're alright, get up and help me round up that tool box and those tools. We gotta get back to town."

"Sure ting, Herol, and . . . tanks alot," Bernie said, his eyes still glazed over with ecstacy.

Back at the hotel they sat at a corner table waiting for Mrs. Geetz to serve coffee. Luckily, the restaurant was empty. Neither Herol or Bernie welcomed company just now. They were in trouble and didn't know what to do about it. Herol knew he'd be patching the wing instead of taking three days off, and Bernie was in hot water for jumping without Carter's approval.

"Hey Bernie, you wanna come with me to the room and maybe help me explain to Bob about this mess I got you in?"

"Me see Tyson? He'd moider me for sure! He already promised me an ass-kickin', remember?"

"Yeah, you're right. I'll talk to him and I'm gonna tell him that I talked you into jumping. How about sticking around the hotel? We'll have supper together."

"Okay, but if dat Tyson is wid you, I'm gonna get lost."

"You ain't been gone very long," Tyson greeted Herol as he entered the room.

He didn't reply but walked to his bed and flopped down in a heap. How do you tell a dumb story like this, he wondered?

"Did you practice anything, or just go joy-riding?" Tyson's mood was gentle — it always was when he wrote letters to Marge.

"Nah, I horsed around a little, me and Bernie."

"You took the squirrel along? Here we go again!" Tyson set the unfinished letter aside and shook his head.

"I boogered up the wing a little," Herol ventured for openers.

"What? Ground loop or something?"

"Nah, the tool box fell through it — and, Bernie did too."

"How in the heck did you manage that? How could the tool box *and* Bernie fall out of the cockpit — Jump? That's it, isn't it?"

"Yeah, I tried to do like when I made my first jump, only I forgot to put Bernie in the rear cockpit, so when I dumped him out the tool box busted through the center section and Bernie got his head stuck in the wing till he got loose." Herol braced himself for the expected barrage.

"I don't believe it!" Tyson roared with laughter.

Herol massaged a couple of day-old bruises on his

This beautifully restored Eaglerock is the proud work of Regan Orman of Dallas, Texas. It is powered by a Curtis OX-5 water-cooled 90 H.P. engine. It is similar to the Eaglerock that Herol flew except that his was powered by a 180 H.P. Hispano-Suiza "Hisso" engine.
(Author's collection)

head for want of something better to do.

"I figured maybe you were pullin' my leg with that first solo in Fargo—the beer, coyotes, and what the devil else—but here we go again! Another 'solo' and this is as nutty as the first one!"

Tyson continued to smile and picked up the unfinished letter to Marge.

"Is Bernie okay?" he asked.

"Yeah, he made a good jump but I gotta patch up the center section and splice some ribs and make a yard or so of fabric repair. The spar's okay though."

"Well kid, there ain't a damned thing I can say that would do any good, except do a good repair job on the wing 'cause I'll look it over. You got three days to do it."

"Guess I'll get with it now." Herol was anxious to get lost in case Tyson's temper flared.

Tyson started to chuckle, "You're a real winner!"

Herol decided that silence was golden, so he tried to smile.

"When you and the squirrel were out screwin' up this afternoon, I got that log book up to date for you. It's all fixed up and it says you soloed—that's the understatement of the century! It will do until this winter when we get down to Fargo and backtrack through the Travalair and Standard log books and get you real legal."

"Ain't you gonna chew me out about all this? It was my fault that Bernie jumped so don't blame him for nothin'!"

"Nah, I ain't gonna chew you out. You're the luckiest dumb bunny or smartest pilot I ever met. I can't figure out the combination."

"I guess me and Bernie were just letting off steam and it got out of hand. I'm sorry."

"Yeah, sure! You're probably the only pilot in the history of aviation that never soloed. That's nuts, but that's you! As for your squirrely friend, I'll tell Carter that you and Bernie were practicing on your own time, so he'll get a pat on the back instead of that kick in the rear he's always expecting."

"That's great! I'm glad we're partners!"

"Sometimes, I got my doubts." Tyson smiled again and handed the log book to Herol. "Wanna eat? — you, me, and Bernie?"

"Bernie too? Swell!"

Herol fingered the pages of his new log book as he followed Tyson out of the room. The homely scrawl of each entry was punctuated with random blobs of ink that were typical of the author's style.

Herol's step faltered as his head collided against the heavy oak door frame. He rubbed an old bruise that he'd just updated to a fresh one.

Then he proudly read Tyson's entry in the log book: Soloed June 7, 1932. 180 Hisso-Alexander Eaglerock A-4, Number C-LVPN. Fairgrounds Racetrack, Coeur d'Alene, Idaho, U.S.A. Robert Tyson Transport No. 2316 Nord Air Flying Circus Saskatoon, Saskatchewan Canada.

A Boeing 80-A Tri-motor. Carter Camp hauled hundreds of passengers in one of these on the air show circuit. Between air show stops, the plane carried spare parts and gas and oil along with emergency provisions. (The Boeing Company — Historical Services)

12 WIN A FEW, LOSE A FEW

The pale moon inched slowly through the morning sky toward the western mountain peaks after a long night's journey. The awakening sun peeked shyly over the crest of the eastern ridge. The cragy peaks split it's rays into pink fingers that reached across the sky over Coeur d'Alene.

A solitary pale light glowed in a window of the Stockman Hotel announcing that Mrs. Geetz was already in her kitchen. Soon the wood stove would be crackling and spewing hot diamonds up the chimney flue. An array of pots and frying pans would play host to the breakfast delicacies that she prepared for "her boys".

Early risers drifted into the dark lobby, greeting each other with a sleepy "howdy" or "good morning". A few wandered into the unlit dining room where the aroma of bubbling coffee and fried bacon drifted through the kitchen door.

This morning was the last day that Mr. and Mrs. Geetz could lavish a generous breakfast fare on their "flier fellers". The gypsy pilots had spent the last two days in feverish activity preparing for their expected departure. Tomorrow's breakfast for them could be only a hurried cup of coffee and their heartfelt goodbyes.

Outside the Stockman Hotel the sheriff pulled his car over to the curb. He was delivering Carter Camp to his troupe of fliers to get them started on their last day in Idaho for this year. Carter was mulling over problems and plans as the car braked to a stop.

"Well Carter, here you are," the sheriff said. "Got lots to do today?"

"Enough," Carter shook his head. "I envy you being able to settle down and stay in one place Sheriff. Must be nice to go home to the same bed every night."

"I suppose," the sheriff mused. "Funny, though, I was just sitting here thinking how much I envy you with your job. Traveling and seeing new places and new people. Not to mention flying them aeroplanes and doing all those crazy things."

"I guess," the sheriff went on, "it all depends on which side of the fence you're on."

"Must be so!" laughed Carter. "See you this evening."

Carter stepped out of the car, slammed the door and walked into the Stockman.

Upstairs, Herol sat up on the edge of his bed and rubbed his head.

"Hey Bob, you wanna eat," He asked?

"Get lost! I don't feel so good. I'll meet you at the fairgrounds later. I gotta look over that wing repair before you sew on new fabric." Tyson rolled over in his snug blanket and resumed his interrupted snooze.

Mrs. Geetz had already doled out ample portions of coffee to her waking boys. They chatted contentedly while she busied herself in the kitchen with the breakfast menu. Mr. Geetz hovered closeby, ready to assist his wife whenever he was needed.

Carter Camp was seated at the large oval table in the center of the dining room. Cups of coffee and overburdened ash trays littered the table. His men crowded around him and listened attentively as he reviewed the numerous chores and responsibilities of each one of them. Today was their last day of preparation. They must be ready for departure by daybreak tomorrow.

Carter's plan was simple. They would take off en masse for Spokane, Washington at first light. They would join up over Coeur d'Alene at a thousand feet in loose formation. He would fly lead plane in the tri-motor. The others would keep their same positions as in the past. This would be a short flight of less than an hour's duration.

Harry Ross had already arranged for the use of meadow land adjacent to a highway east of the city. He'd drawn a map of the landing site and mailed it to Carter. It was now spread out on the table in front of them.

Spare parts that had previously been ordered from 'Pop' Geraud in Saskatoon were available. The heavy crates were at the Spokane freight depot.

Fuel and oil reserves, stored in metal drums were at Spokane. The Boeing would be loaded with this cargo, the spare parts, and emergency rations.

Air show routines would be kept to a minimum so that aircraft could be used for passenger hopping. Without the support of the big tri-motor, the airmen might be hard pressed.

Their departure from Spokane would carry them on a northwesterly course. They would criss-cross the Columbia River at intervals as they traversed the state. Their destination — Vancouver, British Columbia.

Upon arrival, Carter would replenish their lost revenue with dawn to dusk passenger flights in the tri-motor. The others were expected to be equally ambitious.

The men glanced at the doorway when Herol walked in. Carter beckoned the youth to join them at their table.

"Good Mornin'. I was waitin' for Bob. He ain't feelin' good. He's still in bed," Herol said.

"Pull up a chair son. We've been talking about

tomorrow's flight. After you've had your breakfast, look in on Tyson and let me know how he's feeling," Carter replied.

"Okay — Bob said he'd see me at the fairgrounds later on." Herol offered.

"That's fine. Have him get in touch with me when he arrives. I'll fill him in on tomorrow's plans. How's the Eaglerock repairs, finished yet?"

"Bob wants to inspect the wing spar and the rib splices I made before I stitch the new fabric. I'll tell him what you said."

"Guess you've got enough work cut out for you today. Better get some breakfast," Carter smiled.

"Hey boss? Now dat we got da woid, can I eat wid Herol? It's too crowded at dis table for all of us," Bernie chirped.

"Sure, go ahead Bernie. You got all your duties lined up for today?"

"Sure boss — I got da list right here wid me. I'll take care of everythin'. Don't worry."

The group got up from their crowded table and sat down at vacant stools at the counter or tables against the walls.

Herol and Bernie eyed the stools at the end of the counter that would soon be vacated. They stood back and waited patiently for the opportune moment. A friendly grin from the departing cowboys, and they slid into place at the counter.

"Gee! How come Bob's sick? I didn't tink nuttin' ever happened to him. Everybody sez he'll go on forever," Bernie said.

"I don't know. I worry about him 'cause he's so different all the time. I sure like him 'cause he's always looking out for me. He scares the pants off me when he hollers," Herol grinned.

"Yeah, I like him a lot too. All that hollerin' he does used to scare me. He don't mean nuttin' by it. He's the best! — Right?"

"Right!" Herol laughed.

"I gotta drive da bus to Spokane tomorrow when you guys leave. Carter is gonna sell it. What you gotta do?" Bernie asked.

"Fix the Eaglerock — meet you this afternoon?"

"Sure."

Herol and Bernie smiled thankfully when Mrs. Geetz placed two hot dishes of sizzling bacon and eggs in front of them.

"Thanks!" they chorused.

They ate hurriedly. Some of the others were already finished and waiting for Bernie's adventurous trip to the fairgrounds.

Tyson scratched his tousled head and scowled at the walls. He got up on his elbow and reached for a cigarette, then searched aimlesly for a match. He cursed softly under his breath and threw the cigarette to the floor. He sat on the edge of the bed and continued to scowl at the worn carpet beneath his bare feet.

He reached for his head with both hands. The demons in his head were tearing him apart again.

Today was another one of those aimless days he'd been living forever. His thoughts turned to Marge. Would he see her again? Maybe. Could he keep Herol safe until they reached home this fall? What would happen to the kid if he himself didn't survive? He was getting old, wasn't it time to cash in his chips?

The demons scampered off to their never-never land and he sensed the sane world returning. He dressed, went downstairs, and ate breakfast alone.

The ancient bus clanked to an abrupt halt as Bernie ground the transmission gears into reverse. He hopped through the exit to the alley and slammed the loose jointed door shut. He turned the corner and bounded up the front steps of the Stockman, pulled off his glasses, and blew the dust off the thick lenses. He held them up to the light, then gave them a brisk rub on the back of his sleeve. He stepped into the lobby and blinked his eyes for a moment until he became accustomed to the light.

"Hi Mr. Geetz! is any of da guys still around? I'm goin' back to da fairgrounds."

"Oh my yes, Bob is in the dining room. Go right in, he's all alone."

Tyson sat by a window with his back to the door. A cup of coffee was poised in mid-air. A smouldering cigarette dangled from his left hand. He was far away, oblivious to his surroundings.

"Hi Bob, hope you're feelin' better." Bernie greeted him as he bounced into the room.

The unexpected outburst startled Tyson and brought him back to reality. The coffee cup clattered when he set it in the saucer. He turned around to greet his companion.

"Uh — Hi. Pull up a chair."

"Tanks," Bernie said as he sat down.

"You know," a slight smile crossed Tyson's face, "you're okay Bernie."

Bernie's eyes brightened noticeably. The sparse compliment he'd received from this strange man was a priceless treasure. His goggle eyes clouded over a bit as he peered through the thick lenses. For the first time he felt at ease with his complex friend. He even dared to think that he understood. No matter, he loved and respected this giant of the airman's world.

"My goodness! You can't just sit there empty handed," Mrs. Geetz exclaimed. She set a cup and saucer in front of Bernie.

"Let me know if you boys need anything." Mrs. Geetz filled both cups and hurried back to the kitchen.

"You doin' better now, you okay Bob?" he asked.

"Yeah, I guess so. I'll go back with you. I might come back early if you're not busy."

"Sure, anytime." Bernie fidgeted with his coffee cup.

"You've been doing one heck of a good job for all of us. Everybody says so."

Bernie was in Seventh Heaven! He blushed like a dime store light bulb.

"What's Herol doing?"

"He's stiching fabric on da wing. He said he was gonna wait for you before he finished it up."

"Good! Guess we better get out there soon."

"Anytime you say," Bernie replied.

Mrs. Geetz bustled out of the kitchen and refilled their cups.

"There you are, boys. A few hot sips before you go. I'll have an extra nice meal for all of you tonight. I wish you weren't leaving tomorrow," she sighed.

"This is the best home I ever had!" Bernie felt embarrassed with his remark.

Tyson grinned and nodded in agreement. Mrs. Geetz beamed. Bernie was the happiest man on earth!

They sipped the remainder of their coffee in silence. Mrs. Geetz hummed contentedly in the kitchen.

"Ready, Bernie? Let's go!"

They left the dining room, nodding to Mr. Geetz as they went out the front door.

Gimpy Gaines looked out from under the nacelle of number three engine. Carter peered out from the center engine about the same time. The morning silence was broken with a familiar yet muffled sound.

"Sounds like a Hisso engine," Carter ventured.

"Yeah, but where?" Gimpsy asked.

"Damned if I know! Maybe our ears are buzzing from too much flying," Carter grinned.

They scanned the distant horizon and listened, turning their heads in the direction of the elusive sound.

"We must be hearing things. Let's finish checking these engines," Carter said.

Bernie wheeled his ancient chariot past the grandstand. The aircraft were nestled in a grassy area beyond a group of livestock buildings. They were parked at random intervals and formed a rough semi-circle with the big tri-motor in the center.

Tyson strained his senses to catch the sound of a laboring engine. He looked behind him through the rear window of the bus but saw nothing. He shrugged his shoulders.

"You want I should drop you off by the Eaglerock, or you wanna see Carter first?" Bernie asked.

"I'll see Herol first. I'll get to Carter later."

Bernie slowed down and pumped the brake several times in a vain effort to decelerate further. As usual, his tried and true method always worked. He turned the ignition off and the old bus hiccouped and jerked to a stop in front of the Eaglerock.

Tyson shook his head as he stepped to the ground.

"Damn!" was his only comment.

Bernie grinned. He must be improving! He recalled other times when Tyson turned the air blue with profanity and dire threats.

"Hi! Hope you're better." Herol carefully pinned his rib stitching needle into the upper wing and stepped to the ground.

"I'm okay, but I might bail out on you later on." He glanced inquisitively at the upper wing repair.

"How's the wing?"

"Take a look. If you think it's all right, I'll finish the fabric."

Tyson had stepped onto the wingwalk and lifted the partially sewn fabric above his head. He peered intently at spliced ribs and the rear spar. Satisfied with the repair job, he stepped down.

"Looks good! Close 'er up. Do the rib stitching and give the fabric enough clear nitrate dope to pull it taut.

You can finish the taping and doping after we get to Spokane.

"I'll get right on it. By the way, Carter wants to see you." Herol said.

"You're a darned good mechanic, and I'm gonna make you just as good a pilot," Tyson smiled. Herol just stood there grinning from ear to ear.

"You look like a full-fledged idiot! Get busy on that wing!" Tyson growled and turned on his heel, then headed for the tri-motor.

"Yes sir!" Herol called.

Herol was happy. Tyson's harsh exterior didn't hide his gentle spirit within. He turned to the job at hand. There was much to do before nightfall.

"There it is again," Tyson said to himself and looked in the direction of the sound.

He saw Carter and Gimpy in front of the tri-motor. Both were pointing and looking toward the east. They must have heard it too. He quickened his stride to join them.

"Hi Bob — you heard it too!" Carter asked.

"Heard you were sick, glad you're around," Gimpy said.

"Listen! There it is again!" Bob said quietly.

"We been hearing it for the last half hour," Carter said.

They shielded their eyes against the morning sun and stared into the distance.

"Anybody seen Black Jack today?" Carter asked.

They looked at each other for an answer. Then Gimpy volunteered his thoughts.

"I ain't seen him but I do know his airplane ain't here. Ever since Johnnie got killed in Saskatoon, he's been hard at the booze and the broads almost every night. I bet things have just been building up inside him and I figure one of these days he's gonna let loose and do something crazy. Maybe this's the day!"

"Bet you're right! He's probably tearing up the

landscape someplace." Carter replied.

Meanwhile, Tyson had spotted their wayward airman. The sound was nearer and more distinct. A tormented Hisso engine screamed and coughed at irregular intervals. A tiny black speck skimmed along the ground. It slithered from side to side dodging low bushes, rocks, and an occasional tree. It rose a half wingspan above the ground and flipped over to an uncertain path inverted. When the engine sputtered, it tumbled right side up again. The roar grew loader. The scarlet paint flashed in the sun. It was surely Black Jack.

"He's gotta be drunk! I could fly a freight train smoother than he's doing!" Gimpy exclaimed.

"He might get creamed," Tyson muttered.

"Yeah," was all that Carter could add.

The sheriff had driven up in the meantime. He was fond of this strange breed of odd balls. He secretly wished that he could throw convention aside and be the free spirits they were. He certainly didn't want to see one of them killed so he waved at Carter and shouted.

"I'm going to the nearest phone and call an ambulance — just in case."

Carter nodded.

"I'll be back later on if anybody wants to go in town."

A small crowd was congregating near the Boeing. Black Jack's wild gyrations had caught the attention of all those within earshot. They could only surmise what his foggy brain had in store for them. The tiny speck in the distance had grown more distinct but still remained far enough away that it hop-scotched along the horizon or faded into the shadows. The engine bellowed and coughed when Black Jack slammed the wings of the aging Standard to inverted flight.

"What's the damned fool up to?" Carter asked to no one in particular.

"He's my good buddy. He can take care of himself," Dutch Hauptman replied.

"You better take care of your buddy — if you can." Eddie Gardner responded to his wayward mechanic.

"Yeah," a crestfallen Dutch murmured.

"I think he flipped his lid when Johnnie Day got killed. He's had a couple screws loose ever since ." Carter said.

"You're full of crap!" Dutch shouted.

"Shut up!" Eddie shouted right back.

"Let's all get back to work. Black Jack ain't interested in an audience," Carter said.

Black Jack wasn't interested in an audience, but the airplane drew ever closer. Maybe the old Standard was exhausted and its homing instincts longed for the safe haven at the makeshift airport.

The Standard was racing for the center of the fairgrounds. It raced just inches above the rough terrain. Wire wheels kicked up a cyclone of debris as they tore through patches of brush. Rigging wires howled and wooden struts groaned in protest. The tortured wings fought to maintain a semblance of level flight.

Scant attention was paid to the "back to work" order. The demented Standard made a bee line straight for them. It was a veritable threshing machine. A shower of branches, weeds and dirt rained from the wheels and rigging. Black Jack's dark features barely showed above the cockpit edge.

"Hit the dirt!" somebody yelled.

The entire group of men dove for safety as the churning wooden propeller threatened to gobble them up alive. The Standard's nose jerked up a fraction and then the massive wings corkscrewed into a sloppy half roll. The biplane scraped over the cringing bodies on the ground.

"Everybody take cover by the ambulance or bus. Get goin'!" Carter hollered.

The screaming Hisso belched. The biplane sank a fraction and trailed a plume of dust. The nose jerked up and the Standard climbed and rolled right-side-up again.

With the loss of flying speed the nose sank once more and, unable to pull up level in time, brushed the ground and collapsed the left wheel. The engine howled in protest as Black Jack lifted the bruised plane back into the air.

"Wow! He's terrific!" Herol exclaimed.

"Yeah! He's da nuts!" Bernie chimed in.

"He's a goddam idiot, and you clowns are too. He'll get killed soon enough — what's the hurry?" Tyson's eyes drilled into them.

No response — silence was golden!

Black Jack climbed to the comparitively safe altitude of about a hundred feet. He seemed to have the bird under control. The left wheel, axle, and struts swung like a pendulum. He tried several times to shake the mangled structure free, but to no avail. For the moment, he was content to circle overhead until his foggy brain came up with an alternative plan.

The sheriff had set up roadblocks to discourage curious spectators. The ambulance driver was standing by, and a fire truck was on the way. Available fire extinguishers were scavenged from the planes and stowed in the bus.

"What's gonna happen?" Bernie asked.

"I dunno," Herol replied.

Tyson fidgeted in his pockets for a cigarette and cursed softly to himself when he came up empty-handed.

"Black Jack will figure something out — always does," he muttered.

Apparently, he *had* decided something. The Standard slowly climbed downwind until it leveled off at about five hundred feet. A ragged left turn about a mile distant put Black Jack in line with the first pass he'd made. The Standard dipped its nose and bent into a long shallow dive.

"Better take cover — maybe he's gonna try to kill us all!" somebody shouted.

Black Jack squinted down the long nose cowling through the whirling propeller. His goggles were pressed hard against the windshield. Bloodshot eyes were riveted on the target. His aim was just beyond the covey of planes nested on the ground. No longer did the massive wings flounder like a stricken bird. He flew a precise downhill path, straight and true. Black Jack was sobering up. The master pilot was in command. The shattered wheel and axle hung immobile below the fuselage. Rigging wires screamed a soprano pitch as the airspeed stabilized.

"He's gonna do it sure as hell! Get the ambulance and fire truck running! Everybody get in the bus and grab a fire extinguisher. Roll out on my signal!" the sheriff hollered.

The tortured biplane was boresighted on its target. The ground vibrated in unison with the tornadic propeller noise. Black Jack was hardly more than a hundred feet above the ground and racing toward oblivion.

"Let's go!" the sheriff waved his arm.

The biplane covered the sky as it thundred over the tops of the speeding cars. Drivers instinctively ducked, nearly colliding as they swerved off course.

Bang! A dust cloud exploded ahead. Drivers instinctively jammed the accelerators to the floor. Airmen held on to the seats for dear life and squinted into the mushrooming cloud.

Another bang! Tires churned up more dust. Brakes squealed to a stop. Screams of surprise and blue streak profanity rent the morning air. Suddenly the confusion abated. The billowing dust began to dissipate. The scene was coming into focus. The cars were pointed in haphazard directions. Directly ahead, the wooden fence that encircled the race track was demolished for several yards. The sheriff's car appeared to be poised for takeoff.

The red-faced officer slammed the door open and jumped about three feet to the ground. His beautiful patrol car was resting high and dry above ground. The remains of the Standard's landing gear was solidly imbedded beneath the car.

“I must be going nuts!” the sheriff scratched his head.

Sheepish individuals milled around the cars trying to collect their wits.

“I don’t believe it!” somebody blurted.

“I don’t know what the heck is going on,” the sheriff replied.

“Where is Black Jack?”

“What happened to the Standard?”

They faced each other, questioning looks, foolish gestures, nothing made sense.

Finally Tyson assumed command.

“Let’s get in the cars and head back. Black Jack ain’t creamed yet, he’s too smart for that.”

As usual, Tyson’s impersonal attitude unscrambled their senses and stirred them to action.

“Let’s go!”

The sheriff piled in the bus with the airmen and the unlikely caravan raced back across the field.

Black Jack was about five miles away on the downwind side of the field. He’d given the old bird, and himself, a tough workout this morning. The engine was throttled back at slow cruise to conserve his remaining fuel. Vibration from torn fabric, damaged rudder and wings was minimal at the reduced speed.

His rather unusual scheme had worked. He’d counted on enough speed to tear the landing gear loose when he hit the fence. It was a calculated risk. He could have hit the propeller, debris could have smashed the vitals of the plane. He could have crashed to the ground. Anything could have gone wrong.

The Keystone comedy that errupted on the ground was an unexpected tonic for his raw nerves. He roared with laughter — his first good laugh in months.

He couldn’t afford to think about what had happened. He still had to figure what was going to happen next. It was now or never!

He turned carefully into the wind, reducing power and

setting up a long, flat glide.

On the ground, Bernie wrestled his careening bus. Passengers clung to whatever happened to be within reach, including each other.

"What did I tell you? Black Jack can take care of himself!" Dutch shouted.

A chorus of approval resounded through the bus.

The ambulance and fire truck were in hot pursuit and hard pressed to compete with Bernie's galloping chariot.

"Slow down, or we'll all get clobbered!" a shout from the rear.

"Yeah —"

Bernie strongarmed the wheel with his left hand and pointed with the other through the windshield.

"I see him! He's comin' in to land!"

"Get going'!" they shouted.

He stomped on the accelerator, and once more they were off at a gallop.

"Hang on!"

Black Jack was easing his wounded bird down ever so gently. He knew that more wounds were in store. If he could set her down softly enough, perhaps they'd both live to fly another day. He felt guilty for his raunchy behavior this morning. Perhaps the old Standard would forgive him if he decked her all out like new again. They'd been together too long already to part company. Their lifelong love affair wasn't about to end — he'd see to that!

The bouncing caravan continued jolting across the field. Herol was glued to the back of Bernie's seat and hung on to his own as well. Tyson's outstretched legs were anchored beneath the driver's seat. His left arm was wrapped around the sill of the open window.

"Think he'll make it?" Herol asked.

"Dunno! Depends if Lady Luck is smiling today." Tyson was noncommital.

"Gee! I sure hope he does," Herol said.

"Doesn't make much difference. If he makes it, he lives

another day. If he doesn't, he's just another 'Johnnie'. You're born and you die — that's reality. In between is just a second of eternity."

Herol was silent. He'd swallowed many doses of Tyson's strange philosophy before. Although it was a bitter pill, somehow it worked. This harsh concept was Bob's ironclad armour against tragedy.

His short tirade left a sobering effect on the others. Their comic helter-skelter speed run was serious business. The reality was that Black Jack was still in peril.

Maybe a few went along with Tyson's philosophy, but probably, most of them didn't. He was an odd-ball — but so is any genius. He was an exceptional man. Loved or hated — he was always admired.

They were approaching midfield when Bernie started pumping the anemic brakes. He eased the bus to the left and out of line with Black Jack's descent path. The other drivers followed Bernie's lead and closed in on either side of the bus. The three cars pulled up together along the edge and stopped facing midfield.

"This time, let's not goof it up. I guess we're close enough," the sheriff said as he stumbled down the aisle toward the exit.

Bernie gave the door lever a sharp yank. The hinges creaked. The door slammed open with a bang.

"Let 'em all get out — no hurry." Tyson muttered.

Herol did as he was told.

Bernie turned around to face his friends.

"I'm gonna stay right here, I wish dem udder guys would too — but I ain't gonna tell 'em."

"You're on track — to heck with 'em. We'll stay too." Tyson smiled.

"You bet!" Herol joined in.

"Keep the engine running, he's gonna hit in a few seconds," Tyson muttered.

Black Jack sensed a possible wind shift as turbulent air drifted the Standard to the right. He looked for signs of a change — smoke, falling leaves, birds, anything. As a last resort, look for a cow! A million years ago, an old timer had told him, "In a strong wind, cows keep their fannies into the wind. They don't like dust blowing in their eyes while grazing." Could be!

He eased back on the throttle a bit. The big wooden propeller slowed down to a fast idle. The nose crawled up an inch on the horizon. The array of rigging wires and struts hummed a lower key as airspeed and lift fell off.

As the big Standard was poised in a nose high, near stall attitude, the churning propeller tips whipped at tall grass only feet below.

"Let's go!" Tyson shouted.

Bernie stomped on the accelerator, the engine shuddered, then took hold. They were off. The fire truck and ambulance followed in hot pursuit.

A chorus of shouts from all sides drifted through the open windows of the bus as Bernie left them standing in a cloud of dust.

The Standard was hanging on her propeller with the tail inches above the ground. The skid teased tall blades of grass while Black Jack held the plane off 'till the last moment. The turbulence increased near center field. Capricious winds shifted from side to side. The biplane's wingtips threatened to plow furrows in the ground. A wicked quartering wind blasted the side of the fuselage. The sluggish ailerons and rudder were helples. The Hisso roared, the plane staggered and clawed for altitude.

"Turn right! He's gonna go around and land on the short length of the field." Tyson shouted.

The hapless airmen left trailing in the dust were scrambling in all directions. Some ran, others walked, a few stood in the middle of the field.

Black Jack struggled to keep the Standard on an even keel. He glanced nervously at the fuel gauge that stared back at him saying "empty".

Black Jack flew a tight circuit above the field. He was

low and slow — a dangerous combination. He hoped the the thirsty Hisso would keep turning another minute or so. The wind was brisk and choppy. Would it be manageable near the ground! A shallow left turn brought him into the wind. He lowered the nose and slowed the big Hisso down to idle once again. The engine unwound to a soft murmur. He noticed the three cars parked below and to his left. He'd attempt a touchdown as close to them as possible.

Below, the panting runners were conducting a slow motion race to the landing site.

Black Jack was in the groove, the wind was steady. The Hisso backfired, then picked up again. It windmilled a few more turns then stopped dead. He dropped the nose momentarily, flaring out as close to the ground as he dared. The plane bucked. A propeller tip snapped off and sailed under the wing. The bottom of the fuselage dug a furrow, then the plane upended and teetered on its nose and fell back. The dust settled down around him. He lifted the goggles from his eyes, fingered his bloody chin and cursed under his breath. He glanced rapidly at his faithful plane, no major visible damage.

His well-wishers were converging on him. The ambulance driver was running towards him with first aid kit in hand. The firemen were already standing by in readiness in case of fire. The marathon runners were panting out their last breaths. Tyson stood silent beside the cockpit waiting patiently to lend him a hand if needed. Herol and Bernie stood a few feet behind at a respectful distance.

Black Jack hoisted himself out of his seat. His legs were cramped, he was stiff all over. He was too old for this kind of life but now it was time for him to put on a good act.

"Sacre bleau! Ain't you cheap bums gonna applause for the show I just put on?"

He waved at his audience on all sides. A hysterical laugh errupted from his huge chest then he slumped across the cockpit edge in a faint. Tyson cushioned his heavy bulk. The ambulance attendant made a cursory check of his patient, reached in the medical kit and injected a needle in Black Jack's arm.

"Help me get him out" the attendant said.

"What's the score?" Tyson asked flatly.

"I think he's in shock. Best we run him over to the hospital and let the doc take a look at him."

"Somebody get over there and picked up a stretcher," Carter pointed toward the ambulance.

"And everybody get in the bus and stay there 'till it leaves," the sheriff hollered.

The firemen picked up their equipment and made ready to leave.

Black Jack was lifted out of the cockpit and placed in the stretcher. Within seconds he was off to the hospital.

The morning's escapade had wasted valuable time. There was still much to be accomplished before nightfall.

Gimpy and Carter labored silently on the Boeing. Valves must be checked on the three engines. Rocker arm covers needed to be packed with grease. Exterior control cables had to be adjusted for tension. Baggage, miscellaneous tools and spare parts were to be stowed in the cabin.

Herol was confronted with the tedious job sewing the large fabric patch on the wing and rib stitching the center section of the Eaglerock.

Russ Sorensen had the engine cowling removed from the Bellanca and was struggling with a cranky magneto.

'Digger' McGonnigle lent Russ a helping hand. His feisty Travelair was the newest craft on the field and needed little attention other than minor repairs.

Eddie Gardner's Waco was being refitted with an overhauled radiator and water plumbing. Dutch and Pancho stood on the wingwalks and struggled to hang the cumbersome radiator in position between the center section struts. Eddie was busy patching holes made from stone bruises on the underside of the fuselage and tail.

The fate of Black Jack's plane was a problem yet to be resolved.

Bernie tended to the needs of everyone. He collected excess supplies, tools and luggage from each plane and stowed them in the Boeing. He was errand boy or helper for anyone who needed assistance.

"How you doing?" Tyson asked as he sauntered up to the Eaglerock.

"So far so good," Herol replied.

"Tyson eyed the repair job in progress and said nothing.

"You goin' back to town?" Herol ventured.

"Changed my mind — wasn't feeling so hot when I got up and figured I needed to let off a little steam this evening. There's been enough excitement already."

"What we gonna do to-morrow? I don't know what the plans are."

"Just a short hop to Spokane. We'll be there a couple days or so," said Tyson.

"Sounds easy," said Herol.

"The hard job is coming up. It's a long ways to Vancouver! We're heading for the Pacific Ocean!"

"Wow!" whistled Herol.

"Then we're flying north until we see Eskimos!"

"We gonna do all that and get back home before the snow falls?" Herol asked.

"Dunno," Tyson muttered.

"Gee! I might never see Gramps again." Herol was distressed over their bleak future.

"I'll take care of us. I do the thinkin' — remember?"

"Yeah — okay."

Shadows lengthened on the ground, then disappeared as the waning light faded into the western horizon. A palid moon was emerging above the twilight haze and would soon join the night constellations that were awakening.

As promised, Mrs. Geetz had opened her magic book of recipes and prepared "something extra nice" for the airmen's last meal.

Each man had spruced up as best he could. Dusty boots were wiped clean and given a spit polish. Clean shirts and bow ties complimented well scrubbed necks and freshly shaven faces. Tousled hair was neatly parted and slicked down with Vaseline or water. Once grimy hands and fingernails looked reasonably clean. They weren't a very handsome lot, nor would they make the fashion magazines but they were in a class all of their own. And best of all, Mr. and Mrs. Geetz loved each and every one of them!

The dining room had been transformed into a rustic banquet hall. Tables that previously occupied niches along the wall or dark corners of the room had disappeared. They'd been placed end to end in the center of the room. The long table covered with several worn linen cloths could accomodate twenty or more guests. Bouquets of wild flowers nodded to one another along the centerline of the table. Battered silverware sparkled with fresh polish. Ancient plates of unknown origin completed the simple setting.

The large oval table that previously occupied the center of the room was now in a corner. It supported an array of potted plants, cut flowers, and paper wrapped boxes.

Not to be outdone, Mr. Geetz had also risen to the occasion. He was decked out in his black Sunday suit, high celluloid collar, white starched shirt complete with brass collar buttons, a red bow tie — and a big smile!

Mrs. Geetz wore her prettiest pink dress. Her husband liked it because it complimented her beautiful white hair. Her favorite combs nestled in the braids that encircled her head. A flouncy white apron covered the front of her dress. And she might have put a spot of rouge on each cheek; or maybe it was just happiness shining through!

Midst all the happy anticipation, Mrs. Geetz was slightly saddened because one of "her boys" would not be here tonight. She wondered if the doctor might allow someone to take Black Jack his supper to the hospital. Surely, he wasn't that sick! He'd only suffered a cut chin and a fainting spell. Perhaps the sheriff could find out. Although he was on duty, he'd promised to stop in during the evening when he could. Black Jack was her only concern tonight — he couldn't be left out!

Spic and span airmen and a few well scrubbed cowboys lounged self consciously in the lobby. They grinned sheepishly at each other. The dressed-up holiday atmosphere was a rare occasion for these simple, hard-working men.

"Come in boys! Everything is ready! Seat yourself!" Mrs. Geetz clapped her hands.

Tired, happy men filed in through the door. Their eyes bugged out in unison at the transformation of their dining room. It was a sight they'd never dreamed of. They picked their places at random, each waiting for the other to sit down first. Finally, someone scraped a chair against the linoleum floor and they all sat down.

Mrs. Geetz bustled out of the kitchen, her arms laden with two huge baskets of assorted fruit. She placed them near both ends of the table.

"Try all these nice things while I get some more."

Mr. Geetz was close behind her with a huge tin pot.

"Here's coffee for starters," he said as he made his way around the table.

Mounds of mashed potatoes, baked ham, roast beef, and all the trimmings were piled on the table in quick succession. Long arms stretched from all directions as each man attempted to lure a favorite delicacy to his plate.

"This is better than the Waldorf!" Eddie Gardner exclaimed.

"Where's that?" a cowboy asked his sidekick.

"I dunno but I hear tell it's one of them fancy-pants places out east," the knowledgeable cowboy remarked.

"We're sure fancy-pants fellers tonight," his companion chuckled.

Conversation was sparse as the men pitched in. Mrs. Geetz made sure that everyone was too busy eating to socialize.

The telephone out in the lobby jangled and brought Mr. Geetz out of the kitchen. He hurried across the dining room and into the lobby.

"Hello? Oh yes, sheriff. He is! Oh no! Oh my! We shouldn't? You will? Bye!"

Mr. Geetz wiped his forehead with a big handerchief, and hurried back to the dining room.

"Hey fellers, the sheriff just called. As you know, he rode back to town in the ambulance. Black Jack is Okay — but he isn't."

"What's wrong?" a chorus of voices inquired.

"Nothin' is wrong — or maybe it is. Oh my! I'm all confused!" Mr. Geetz was getting nervous.

"Maybe we ought to get down to the hospital after we eat." Carter Camp suggested.

"Oh no! The sheriff said we shouldn't come! He'll be here later, he's busy now." Mr. Geetz said.

"Sounds screwy to me," Carter replied.

"It don't sound screwy to me," Tyson muttered.

"What do you mean?"

Mrs. Geetz came to the rescue of her husband and broke the impasse. She bustled around the table and placed a little gift by each man's place. And then she walked over to the side of her husband and smiled.

"We got a present for Black Jack too," Mr. Geetz said.

"HOORAY FOR MR. AND MRS. GEETZ," they shouted.

"Open your gifts — they're not much. We hope you like them." Mrs. Geetz said.

"Hot dog!" a cowboy discovered a tiny rocking horse.

"Hey look!" His buddy held up a miniature hammer.

"That's for rebuilding Gimpy's leg," said Mrs. Geetz.

Dutch and Pancho got toy tool kits. Russ, Gimpy and Eddie had little wooden airplanes. Bob and Herol received toy knights — one black the other white.

Carter's gift was a little sign that said "Boss". Bernie got one that said "Official". Mr. and Mrs. Geetz still had Black Jack's gift.

Mrs. Geetz went over to the corner of the room and opened the lid of her lovely old gramophone. She put on a record, placed the needle in position and wound up the machine. A beautiful Viennese waltz began floating across the room in slow motion until the record picked up speed.

"Come on Papa — you haven't asked me to dance in a long time."

"May I have this dance, Mama?" Mr. Geetz responded.

"I'd be delighted!"

The old couple glided across the floor with the beauty and grace of bygone days.

Mrs. Geetz didn't realize what she was in for — or maybe she knew all along! There were willing hands to keep the gramophone wound up, and eager dancers who waited their turn with the sweet old lady.

"IS HE HERE?" A wild eyed sheriff burst into the room.

"What! Who?" Mr. Geetz stuttered as he confronted the sheriff.

"That crazy Black Jack! Who else?"

"I knew it!" Tyson chuckled.

"He threatened to tear the hospital apart! The doc wanted him to stay overnight, but he had other ideas."

"Oh my! Oh my!" was all Mr. Geetz could say.

"Please calm down, sheriff," Mrs. Geetz interrupted. "Come sit down and we can talk about it."

"Sorry, ma'm — I gotta track him down. He's on the loose!"

Mr. Geetz was glad that his wife was here to handle the situation. All that he could think to say was " Oh my!"

"Now sheriff, you said you were coming over this evening, and here you are! I'm sure Mr. Black Jack knows what he's doing."

"He's a sick man," the sheriff growled.

"He certainly doesn't sound very sick to me. He's just worried about his airplane."

"Well — uh — I'll stay a couple minutes."

Mr. Geetz hurried over with a cup of coffee. The sheriff sat down and tried to unravel his frazzled nerves.

"Hooray for Black Jack! I knew he'd be all right!" Dutch shouted.

"What are you going to do about your buddy?" Eddie Gardner asked.

"Shucks! Nothing, I guess. He sure ain't goin' with us tomorrow."

"Hey Carter! Would you come over here a minute? We need your advice." Eddie called.

"Sure — what about?" Carter asked as he approached.

"Since Black Jack and his plane are both bent out of shape, I thought maybe we could help."

Eddie glanced at Dutch and hoped that he might pick up the idea.

"How about I stay here and help Black Jack? That is, if it's okay with Eddie. We could meet you later on the road, or back in Sasketoon."

"Sounds reasonable," Eddie said.

"We've still got Pancho to help us."

"Sounds good to me," Carter replied.

"It's up to Black Jack if he wants to repair his old turkey or try to pick up another. I could still use him on the Boeing or the Bellanca after Russ leaves."

"Come on sheriff! You're the only one I haven't danced with tonight." Mrs. Geetz pulled the sheriff off his chair.

"Ah shucks ma'm, I'm on duty."

"No excuse! Come on!"

For the first time today the sheriff felt relaxed. It had been a rough day. His expertise as a dancer was limited to a few basic steps but he wasn't bashful. Mrs. Geetz was light as a feather and anticipated his every move. They swept across the floor in perfect unison. The sheriff felt wonderful!

"HI EVERYBODY! — I got delayed." Black Jack burst through the door.

His massive bulk seemed to fill the room. His attire was hardly up to the dress code that prevailed here tonight. An ill-fitting hospital gown was tucked into his pants. His massive boots had been replaced by a pair of sick room slippers. He sported several dark bruises on his face and three stitches in his lip. Black Jack had his own dress code tonight!

Once again Mrs. Geetz was left deserted on the dance floor. The sheriff stomped over to Black Jack.

"I been lookin' for you!"

"On the dance floor? I get the next dance with Mrs. Geetz! A gentleman wouldn't leave a lady stranded!" Black Jack smiled his angelic best.

"May I, Mrs. Geetz?"

"I'd be delighted!"

"I don't believe it! I don't believe it!" the sheriff sputtered.

A chorus of friendly laughter rose through the room.

"You win a few. You lose a few," the sheriff chuckled.

"You ain't won any today!" Bernie chirped.

They all roared!

"Mr. Geetz, may I use your phone? I better call the hospital," the sheriff asked.

"Oh my yes! You know where it is."

"I better talk to the doc — thanks!"

Black Jack and Mrs. Geetz danced around the room until the music finally ground to a stop. The gramophone had run down, and so had Mrs. Geetz. Black Jack escorted the tired lady to a chair. He thanked her and made a gallant bow.

"Don't you go away! — Papa! Give Black Jack his gift."

"Oh my yes!" Mr. Geetz handed him the little package.

Black Jack's battered features softened into the semblance of a smile. The boisterous, hard fighting brawler was a little boy for the moment. He tore off the plain brown paper wrapping and there was a tiny silk parachute.

"That's your parachute Mr. Black Jack!" said Mrs. Geetz.

Black Jack unfolded the silk cloth. The ends were tied with thread to a small toy figure.

Gently he swung his hand back and then up. The little silk ball arched up toward the ceiling and as it turned to come back down, it opened with a soft 'plop'. The parachute floated down gently and Black Jack put out his hand and caught it.

Hands clapped! Shouts of Hooray! Hot dog! And more clapping with shouts of "Speech! Speech!"

This big giant of a man, who'd captivated audiences for over a decade had finally met his match. This audience was the best ever, his sky-happy buddies, his cowboy friends, and the two gentle hearts that made this night possible.

He stood in the middle of the floor and looked around at the roomful of smiling faces. He was scared to death! He had stage fright! It couldn't be!

"Uh — thanks to everybody. I ain't ever owned a 'parachute' before. This one I'm wearing every day of my life — right next to my heart!

The big man was visibly shaken. He stumbled over to a chair and sat down. He was a happy man!

Pancho had quietly stepped out of the room and reverted to his native costume. It took little imagination to envision a 'south of the border gaucho' as he reappeared. The Mexican street urchin was resplendent in a sparkling bolero jacket, a matching sash at his waist, and a colorful bandana atop his dark hair. Black eyes flashed as he improvised a sensuous latin rhythm on his mandolin. He stepped softly around the room, keeping time with the exotic beat.

Tyson rose from his chair and politely asked Mrs. Geetz to dance.

"Oh goodness! I don't know them beautiful Spanish dances," she exclaimed.

"You will!"

They stepped to the center of the floor and he guided her into a graceful bolero. Pancho smiled with approval as he watched their beautiful interpretation of his music. Tyson was an expert dancer. His strong lead gave Mrs. Geetz the confidence she needed. Soon, she too, was carried away with the haunting strains. Tyson's black attire complimented her beautiful white hair and pink gown. They were poetry in motion!

Once again, Tyson had unwittingly revealed another facet of his strange being.

He escorted her back to her chair midst the applause of their appreciative audience. Beauty and grace seldom touched the lives of these rough-shod men.

The sheriff had finished his phone call and was lounging against the dining room door. He'd been equally entranced with Pancho's stirring music and Tyson's natural grace. His crusty heart went out to these vagabonds who dared to be different. He looked across the room at these men. Each one was an individualist, yet so alike. There was a fierce loyalty among them. It buffered them against the daily odds they faced. He wished they weren't leaving. Life would be empty. He decided that the rules could be bent one more time in Black Jack's favor.

He stepped away from the door, took a few steps, and seated himself next to Black Jack.

"I talked to the doc. He wants you back at the hospital. He wants to keep an eye on you for a few more hours."

"Ah! I don't need to go back. I ain't ever been sick."

"You got a choice — the hospital or jail!"

"What!"

"Yeah! That's right! — disturbing the peace and wrecking public property. You got a bill from the ambulance driver, the doc, and what else. I don't know!"

"Oh! I can pay all that off — and then some. I'm sure as heck sorry for giving you the run around all day."

"You win a few — you lose a few," the sheriff repeated his pet saying.

"I guess you're winning this one," Black Jack said.

"Mr. Pancho! We thank you for your lovely music. Would you do me the honor of dancing a waltz with me?" Mrs. Geetz smiled.

"Si! I will teach you the Mexican version."

A cowboy cranked the gramophone and they were off across the floor. Pancho's style was as beautiful as his music. Mrs. Geetz danced a latin rendition of the Viennese waltz.

The airmen were showing signs of fatigue. It had been a long day. It was time to say goodnite, and soon it would be time for goodbyes.

The sheriff made the first move.

"I been here too long folks. I gotta tend to business. It's hard leaving such good company."

Black Jack rose alongside the officer.

"I gotta go too, see you in the morning before you leave — I hope."

They said their goodbyes and strode through the front door and into the night.

"Where to sheriff?"

"Where you want to go?"

"The hospital. Let's go!" laughed Black Jack.

"Right!" the sheriff said smiling.

**A Bellanca Pacemaker. Russ Sorensen flew a Pacemaker for Herol's airshow outfit. It was mainly used to haul passengers for money during the airshows.
(Author's collection)**

13 THE LAST BULLET

Amber skies embraced chilly winds and gave promise of an early winter. Frothy water chopped against the Pacific shore a hundred yards from the sand dunes where Happy Bottom's gypsies were encamped.

From Coeur d'Alene, Idaho the Nord Flying Circus had made the short hop to Spokane, Washington where they'd taken on fresh supplies and some new equipment. Newly provisioned, they headed across the Washington prairie hitting Cheney, Moses Lake, Ephrata and Wenatchee.

After the air shows in Wenatchee they threaded their way through the Cascade mountains and emerged at Everett on Puget Sound.

At Everett they turned north and flew up the coast of the inland waterway into Canada and on to Vancouver.

They had stayed at Vancouver several days, flying air shows and passengers from almost dawn till dusk. And when Carter Camp figured they'd done about all they could do there, they packed up and flew on up the coast stopping at Powell River, Campbell River, Port Hardy where they finally saw the Pacific.

From there they'd wandered northwesterly to where they were now, the remote village of Prince Rupert, British Columbia, a stones throw from the southern tip of Alaska.

Summer had taken its leave for another year and autumn was on the wane. They should have been in Saskatoon weeks ago for a well-earned rest, but each week's delay jingled more money into the company coffers and a fatter bundle for everyone to tide them over through the winter ahead. Days merged into weeks and falling leaves foretold the approach of snow. They would be lucky if they beat the early blizzards in the high elevations of the Canadian Rockies.

Carter Camp had done a fine job as usual. The group he shepherded now was a crack team of skilled daredevils that played for the highest stakes, yet accepted discipline from him that would have been scoffed at last spring. No longer were these men "aerial saddle tramps," as Happy Bottom referred to them, but men of courage with a new lease on life. They didn't affect the bravado of yesteryear to hide their hungry guts. Instead, they were reserved, proud men. Even their appearance reflected this changed attitude. They sported new boots, breeches, rawhide leather jackets, and helmets. At first glance, they might have been mistaken for a team of socialite polo players.

A few would not make the return trip this year. Already the ranks were thinning as some went their separate ways in search of greener pastures.

The first to leave was Russ Sorenson, who said good-bye to his friends when they reached Vancouver. His sojourn was over until next year. It was time for him to return to the States and resume his airmail run from Los Angeles to Salt Lake City.

One day Black Jack simply disappeared with his buddy, Dutch Hauptmann, and they were never seen or heard of again. Everybody guessed that they'd followed Johnnie Day's footsteps and headed for South America to join up with any of the revolutions that were forever erupting.

"Digger" McGonnigle volunteered to stay with his friends until they reached Saskatoon, although he would leave them then to fly southwest to Los Angeles to meet Russ. They had become good buddies as they shared the Canadian skies. Russ enjoyed hearing the make-believe heroics that Digger was so fond of inventing, but never believed a word of them. It must have been Digger's boyish exuberance and flying potential that won him over. At any rate, Digger intended to join Russ in a few weeks and try his skill as a fledgling airmail pilot.

Bernie Tannenbaum was with the only friends he had ever known, and would stay with them as long as he was needed. No longer was he the wisecracking guy from Brooklyn who knew everything about everything, but a seasoned veteran in his own right. He could hustle the suckers through the ticket gates to the tune of several hundred dollars an hour, and his proudest moment was when Carter allowed him to make an occasional parachute jump. Bernie's showmanship was far less than spectacular. But he filled in the voids when stuntmen were too exhausted or nursing broken bones.

"Gimpy" would also return to Saskatoon to collect his season's wages, then he'd leave for Winnipeg where a job as operations manager for an embryo airmail service awaited him.

Tyson was still regarded as the top stuntman of the day. "Diablo" had no intention of being surpassed, and as far as anyone knew this strange character would go on forever. He would return to Saskatoon, flying the Bellanca home in Russ's absence. — And there was Marge.

Herol would follow close on the heels of his partner and bring the Eaglerock home.

Eddie Gardner and Pancho were the remainder of the small group that would traverse the hundreds of miles of wilderness back to home base.

Back in Saskatoon, Happy Bottom scratched his backside absently as he nodded approvingly at the column of figures, gate receipts sent back by Carter Camp, that represented an all-time bonanza for his flying circus. After eight years, his hunch was paying off. Thanks to Carter, the numerous accidents he'd expected were minor. Best of all, no big fines were paid to bail his gypsies out of the clutches of the law. He'd reap the lion's share of the net receipts, but the airmen too would see more folding money this year than they had expected or ever dreamed of. It would be the horn of plenty for all. He was anxious to see his boys safely home.

Miss Savageau sat in her outer office reflecting over the events of the preceding months. A week never passed that she didn't hear from Tyson. Sometimes it was a card, more frequently a long letter written in pencil and punctuated with erasures, mistakes, and the homely language that only he could write.

The strange attraction that she felt for this man still frightened her, but in her heart she sensed the gentleness he was so reluctant to show. An unexplained dread always lingered, not the fear of death from his crazy lifestyle, but the tortured soul that haunted his eyes, crying for release.

He had written many times of quitting the air shows and settling down to a more prosaic flying job, and hoped that maybe Happy Bottom would offer him a year round job. Between the lines, he'd spoken of marriage, although he was too shy to ask her. She'd say yes if he ever mustered up the courage to ask her, and if he didn't, she'd help him.

Herol pushed aside the mosquito netting from the entrance of the tent that he and Tyson shared. He'd put the Eaglerock to bed for the night after a grueling day, and as he entered the tent he dropped the bundle of silk from his chute on the cot. He groaned as he gingerly fingered the bones in his neck.

When his vision adjusted to the darkness of the interior, he saw Tyson sprawled on the other cot where he had been since morning. It was another day of brooding and Herol knew that he was in for a tail twisting for something or other.

"You dumb squirrel! You jumped this evening, didn't you?" Tyson sprang to a rigid sitting position and glowered at his partner.

"Yeah, I got worn out hopping passengers all day. I musta made two hundred landings in that sand pile out there. I was supposed to be straight man for Gimpy, but his wooden leg was busted or somethin', so, I asked Eddie Gardner to fly for me and I made a drop for Gimpy."

"You ain't got a brain in your head yet, and I figured Eddie was smarter than to be took in by you! I told you a long time ago, I'd do all the thinking for us — now you're trying to get yourself mashed in a jump." Tyson's cold eyes drilled into him.

"Ah! I ain't jumped, or nothin' for a couple months and you don't show up half the time no more. All you do is write letters and I gotta fly all day." Herol regretted the rebuke he'd given his friend, but, as usual, it was too late to make amends.

"By damn! I got reasons for doin' things for both of us, and I'm gonna keep you alive till we get back home. If you want to try for a busted head this winter when I'm not around, that' okay. You want a busted head, or you want to fly a few more years?"

"I want to fly forever!"

Tyson relaxed a bit, and in a more gentle tone, "You're gonna fly plenty — and no more of that stuntman crap, maybe never again! I been trying to make a darned good pilot out of you like I promised Cash Chambers back in Winnipeg. You better forget about this grandstandin' and hero stuff."

Herol decided that he best remain silent. Tyson's submerged wrath could still surface.

"How long you been wearing that horsecollar around your neck?" Tyson asked in a matter of fact tone.

"About two months, I guess."

"You're dumb for sure! I saw you rubbing your neck when you came in. Why didn't you tell me you was gonna jump?"

"I was gonna tell you — I figured one jump wouldn't hurt."

"I shouldn't even let you fly with that cracked neck, but I figured there wasn't all that time to spare, so I been watching you and took a chance you wouldn't get hurt any more. You're lucky you didn't come floating down dead as a mackerel. All you needed was a hard jolt when that chute popped and you would have fell through your fanny." Tyson turned gentle.

"Ah! I'm sorry! Guess I'm dumb like you always tell me. I'll get mashed sooner or later if you ain't around."

"No more grandstandin' for you — you're a lousy ham anyway because you believe all that hero crap. From now on, you're gonna do nothin' but fly. I ain't gonna do a thing but keep my eye on you. Gimpy and Bernie can do all the parachuting. Whenever your neck bothers you, I'll fly for a spell. Okay?"

"Okay." Herol felt relieved, somewhat smarter, and chastised as usual.

"Let's pack your chute and find some grub — you hungry?" Tyson smiled as he grabbed an armful of silk.

"Hungry? I could eat a bear!"

Carter pondered over a World Atlas and scores of road maps as he considered possible routes that would bring them home. The mountain blizzards worried him, and he admitted to himself that the lure of money had warped his better judgment. They should have departed weeks ago, but it was too late for regrets. The shortest route from Prince Rupert would be southeast across British Columbia. They'd enter the wilderness and feel their way through the mountain passes and hope for safe landing areas. He knew the wilderness awaiting them would claim the hapless victim who made the slightest error in judgment. Perhaps their major stop would be Edmondton, Alberta, a city that boasted a fine grass airport and probably the only one they'd find across the continent until they arrived home.

He decided that tomorrow they would begin their long journey home. Gimpy would fly copilot in the Boeing with him, and Harry Ross and Bernie could deadhead as passengers. He was going to have Tyson fly the Bellanca, and Herol would follow in the Eaglerock. There remained Digger, who would fly his Travelair, and Eddie and Pancho in their Waco. Only five planes and nine men would be returning to Saskatoon.

It had been two days since when the Nord Air Flying Circus had departed Prince Rupert. It was afternoon now.

Tyson strained uncomfortably in the cockpit of the Bellanca as he tried in vain to see through the ice-coated windshield. He relished the comfort of the enclosed cabin, but somehow felt out of place without the propeller blast in his face and hearing the song of flying wires in his ears.

He glanced over his left shoulder to see Herol rubbing the ice from his goggles as he inched the wings of the biplane closer to the flat underside of the Bellanca's wing. They flew a tight formation through the howling blizzard, ever mindful that if one lost sight of the other they would collide in the blinding snow.

They traced a zigzag course through the narrow canyon that meandered between sheer cliffs that towered hundreds of feet above them. Tyson kept a wary vigilance to his right. The dark blur came into sharp focus as rocky ledges reached out for the fragile craft. Tyson knew that if he didn't outguess the granite walls so near his wingtip, the kid would fly to his death too. If they didn't make it, the explosions from their planes might alert Carter, who was somewhere behind them in the ungainly transport.

Digger and Eddie were tucked close together in formation. Digger's feisty Travelair played cat and mouse with the canyon walls as Eddie flew blindly alongside and struggled to keep his slower Waco in position off Digger's left wingtip.

Pancho's head moved from side to side in the front cockpit of the Waco. He was terrorized by the frigid world surrounding him and thought wistfully of a long time ago when he was a street urchin in sunny Mexico.

Tyson wondered how far behind them an explosion could be heard in the event this canyon led them to a dead end. Maybe Digger or Carter would hear the impact in time to crash land in the wilderness below. Their odds for survival were pretty grim, but hadn't they been like that for months?

Carter's forehead was covered with beads of sweat despite the freezing cockpit temperature. He'd wrestled the Boeing's heavy yoke for what seemed hours to maintain a semblance of level flight in the gusty air. The white glare outside was magnified by the ice crystals that sparkled on the windshield.

Gimpy shivered in the copilot seat as he tried to see through the half-open window on his right. The wings, struts, and wires were encased in a frothy coat of ice that frequently broke off in a noisy clatter. Showers of cutting knife edges banged against the fabric covered fuselage and disintegrated against the tail.

Harry Ross was glued to his wicker chair far back in the cabin. For the first time in his life he didn't have a word to say.

Bernie squirmed or ducked as the avalanche of ice riccocheted off the window pane beside him. Bernie wasn't talking either.

Carter could only guess where they were and how far behind they might be from the Bellanca and the Eaglerock. Forward visibility was nil, and it was possible that they might be overtaking their companions ahead.

Gimpy kept his vigil and guided Carter to the left or right with hand signals as the rocky walls slid past them. He thought briefly of the old days when he flew the mail.

They were in a hairy fix just like old times— but it still beat working!

Miss Savageau sorted through the morning mail and daydreamed how nice it would be if Bob and his young companion walked through the doorway this very minute. She noted that Carter's weekly report was missing from the stack of letters on her desk. In all probability they were on their way home and somewhere in the midst of the vast wilderness to the west.

Happy Bottom's shrewd insight told him that it was time for a change. The financial world was going through a period of cautious optimism as the throes of the long depression lessened and new ventures were underway.

The Provincial government as well as the Federal government in the States were knuckling down harder than ever to restrict unsafe practices in the transportation field. He'd heard rumors that the days of the flying circuses were numbered, and the carefree barnstormers would be required to show their proficiency in order to qualify for a license. Their aircraft would be inspected as well to determine their air-worthiness. He knew his saddle tramps would rebel at government restrictions. They'd been free spirits too long and had already proven their skill by the sheer act of survival.

It was the opportune time to progress from the sideshow aspect of flying to the professional business world of aviation. The north country might be ready for a newer mode of transportation than the railroads. The trains were locked to the steel tracks that took years to push through the wilderness, whereas the plane was as free as the pilot who flew it and would crisscross the continent in hours.

He made a mental note to ask Miss Savageau to acquire all available government information on commercial aircraft operations. He'd discuss this at length with Carter who should be returning any day.

The skies opened momentarily as the blizzard lost momentum and the harsh winds subsided. An occasional patch of blue far above gave promise that the worst of the storm was over. As its fury abated, layers of thin strata engulfed the mountain tops and spilled across the drifting snow in the canyons below. The frigid temperature had risen a few degrees and the powdery snow turned into a wet sticky mess that clung, then froze to the over-burdened aircraft in a thick blanket of corrugated ice.

Eddie's Waco staggered against the new onslaught of ice. He lagged ever further behind Digger, whose agile Travelair sported twice the horsepower of his cantankerous Curtis engine. He poked his head into the cockpit at frequent intervals as he watched the needle of the water temperature gauge creep near the boiling point. The radiator was being strangled for cool air as the sticky snow built a barrier of ice across the radiator core. Steam trailed across the top of the center section wing and above Eddie's head, reminding him that the life's blood of the engine was bleeding away. If the weather didn't change soon, it would be curtains for him and Pancho.

They'd lagged so far begind that Digger disappeared in the layers of strata ahead. The engine coughed as it gulped a slug of watery gas then resumed its loud bark. Eddie pushed hard against the throttle, hoping to coax another revolution or two from the engine as they began to stagger at a nose high angle to fight the persistent loss of altitude as they sank down the canyon walls. Eddie reasoned that Tyson and Herol probably passed him overhead. The several hundred feet of altitude that he lost probably averted a collision, so it was another slim lease on life for all of them — whatever that was worth. He couldn't wait any longer. The engine was running hotter by the minute. He reached down into the right side of the cockpit floorboards and pulled the fire extinguisher from its bracket. He leaned forward across the windshield and tapped Pancho on the head with his gloved fist, handed him the extinguisher, and pointed to the radiator. Pancho would get rid of that ice with the only tool available to them. He had never been inclined to venture beyond the safe confines of the cockpit like the rest of those fools, but this was no time to argue.

Digger was miles ahead of Eddie and Pancho. He knew they were in difficulty and could only hope they were still in flight. He felt secure in his Travelair even though the big Hisso was running hot and coughing from the ice it had swallowed. He'd been through worse storms before and made it, and the weather was lifting. The radiator was running too hot but it seemed to be holding its own. He tried to move the shutters and cursed when he realized they were frozen shut. He skirted between thin layers of strata and grinned when he saw the lower layer disappearing. He caught brief glimpses of trees below, and sparkling snow on the ground promised sunshine ahead.

Tyson relaxed his vigil a bit as the blowing snow diminished and the canyon came into sharp focus. A thick layer of fog floated above the top of his wing and merged into the shadowy horizon ahead. He lowered the nose of the Bellanca to descend into warmer air and sank through the light snow showers until the valley was visible. He glanced over his shoulder to see the Eaglerock less than ten feet away, tucked snugly in position under his wing. The radiator on the nose of the Eaglerock was almost close enough to touch, and he could see eddies of steam spewing from the side louvers of the nose cowling. The big Hisso was running hot but the kid was handling it okay — he saw the nose shutters open and close several times to break away ice accumulation. The Eaglerock's long wingspan carried a rough coating of ice on the leading edges and heavy accumulations on the interplane struts. The flying wires vibrated as slivers of ice broke loose and tumbled past Herol's head.

Tyson noted with satisfaction the small patches of sunlight on the ground below and decided to continue their descent to treetop level. The warmer air would rid them of most of the ice, and Herol's engine would get a chance to cool down. They eased away from the rock wall and held a course that would take them about centerline over the canyon floor. As snow flurries lessened, he guessed their approximate distance from either wall by the dark shadows that still lurked in the snow clouds on either side.

They raced past scrub timber at ground level, instinctively climbing when they approached a clearing. Snow blindness would temporarily overcome them as they sped through the white world. The ice turned to mush and slithered off the Bellanca's windshield, and Tyson breathed easy for a change knowing the worst was over. He could see the Eaglerock trailing behind at a comfortable interval and felt confident that Herol's engine was roaring a healthy tune.

Herol hadn't seen a thing except the underside of the Bellanca's wing for over an hour. Now it felt good to fly an easy trail position behind Bob and give his nerves a chance to unravel. His boots were dead weight on the rudder pedals and he was sure that his feet were frozen. He touched an icy mitten against his face and felt nothing. He knew for certain that his goggles were frozen to his cheek bones. He grimaced as the last sliver of ice somersaulted past him, then pushed the throttle ahead to catch up with his companion.

Digger cursed as the smell of burning oil filled the cockpit. He stretched his neck past the windshield and saw black foam oozing through the engine louvers and stretching long fingers back toward the cockpit. He ducked behind the windshield and studied the engine instruments. The oil pressure was down ten pounds and the water temperature was climbing fast. The engine was laboring from excess heat and it was only a question of time when all the oil foamed out of the tank or the radiator burst. He reduced power to a fast idle and descended through the broken deck of clouds beneath him. He cursed again because he hadn't let down to a lower altitude before now. If he was lucky, the engine would hold out until the ice melted.

Carter and Gimpy shielded their eyes from the expanse of bright snow below. Patches of blue intermingled with yellow sunlight blinded them until their weary eyes became accustomed to the welcome light. They'd inched their way down cautiously through the clouds for the past half hour, feeling for the canyon floor, ever aware that their ungainly craft would respond too slowly if they had to level off quickly. The ice melted from the windshield during the descent and the most stubborn remnants on the wings were peeling off and tumbling past the cabin windows.

The silhouette of a biplane popped in and out of view as it disappeared into the pine shadows, then reappeared against the backdrop of blue sky. Carter was glad to see they weren't completely alone. Now he knew that at least two of them stood a chance of getting through. He eased the throttles ahead to join up with the unknown craft. It could be Digger, Eddie, or Herol. He fervently prayed that he'd see all of them soon.

The mountainous terrain melted into rolling hills that were covered with green pine and dense maple forests. Small clearings punctuated the timberland where itinerant logging camps had wintered. A frozen river divided the cleared land indicating that logs once floated downstream to a mill or possibly a small community.

Pancho slouched low as possible in the Waco's cockpit to avoid the frigid gale that swept past him. He beat his arms across his chest trying to restore the circulation in his chilled body. At the expense of frozen hands, he'd saved the day for them. The Waco purred contentedly, free of the strangling ice. They flew at ground level and rose or fell as they followed the hill contours. Pancho poked his head out, turned around and grinned at Eddie, then looked up and pointed excitedly behind them.

Carter and Gimpy grinned back at him and inched the big bird a little closer. Pancho waved excitedly, and Eddie turned his head and nodded. He noted Bernie and Harry waving from the cabin of the Boeing and waved back.

They hopscotched from tree tops to clearings and down river beds, savoring the warmth of the sun that revived their cold-soaked bodies. They zoomed over a snow-covered hill and slid down the other side to find themselves trailing Herol and Tyson by a mere hundred yards. They joined up in loose formation as they followed a river downstream, confident that eventually they'd find a settlement and a possible landing sight.

Carter tried to convey his thoughts about Digger to Eddie by using hand signals. Eddie shrugged his shoulder in response. He couldn't understand or he didn't know. They flew on for several minutes, each man feeling more secure in the other's company. If only Digger would join up, then they'd all be together.

They climbed along the slope of a hill until their wings grazed the low hanging clouds at the summit, then slid down into the valley several hundred feet below. About a mile ahead, Digger circled above a clearing. The Travelair trailed a mixture of oily smoke and white vapor behind as Digger glided silently down for a forced landing. The engine had finally given up, and he knew that he was completely on his own.

Herol slid away from the Bellanca and down toward the clearing, intent on landing behind Digger. He'd have him back in the blue before he had a chance to get his boots wet.

In the meantime, Digger sideslipped over tall pines that rimmed the field and was settling in for a landing. The biplane skidded when the wheels lost traction and the tail rose high as deep snow took hold. The broken propeller churned a furrow into the snow as it skidded to a stop. The plane teetered in the wind, then fell back on its tail. Digger scrambled halfway out of the cockpit and waved, signaling that he was unhurt.

"That dumb kid is gonna cream himself in spite of me! But I'll get him home alive if it kills me!" Tyson nosed the Bellanca down and flew a beeline toward the Eaglerock's path. Herol was intent on stalling the plane in as slowly as possible so that his landing roll would be reasonably safe in the deep snow. He'd get them both out of here in nothing flat! A shadow crossed in front of him as the Bellanca loomed big as a barn and blocked his path. He slammed the throttle forward and stood the Eaglerock on her wingtips to avoid hitting the Bellanca. Then he staggered across the field and clawed for altitude. He shook his fist, and cursed a blue streak at anyone that cared to see or listen. When he cooled down enough, he joined his companions who circled overhead.

"Maybe it just wasn't a good idea after all," he said to no one in particular. "But what the heck is gonna happen to Digger?" He decided to follow the Waco ahead of him and let the older guys decide what to do next.

Digger sat on top of the fuselage with his boots dangling in the cockpit. He continued to wave with his left hand while he brandished a .45 caliber automatic in his right. He couldn't stop waving because he didn't know what else to do. Maybe it was goodbye, and maybe he was scared as hell!

It took little imagination to realize that he had every reason to be afraid. A pack of timber wolves that skulked in the forest shadows mustered enough courage to venture into the open and circle the downed pilot. The aircraft flying overhead frightened them at first, but they soon realized there was no danger. The pack leader tightened the circle and howled as he caught the human scent downwind.

Digger didn't wave anymore. The goodbyes were over and kill or be killed was the task at hand. He held the gun poised to fend off any sudden attack and shouted obscenities at the approaching beasts. They slowed their encirclement briefly, knowing that strength in numbers was to their advantage. Digger fired point blank at a ravenous animal that broke ranks and leaped for his throat. The .45 slug tore a hole in the wolf's skull. It danced in mid air, then fell dead in a heap. Digger waved again as if to assure himself that he was still the victor.

The treacherous skies of a few hours past felt like a safe haven as each man viewed the struggle that took place below. They fantasized impossible schemes to rescue Digger, then hopelessly realized there was nothing they could do. They continued their watch and frequently one of them would race across the field in a vain attempt to chase the wolves into the forest. Even this proved to be of no avail. The wily animals ran in relays — half of them toward the woods while the others attacked Digger.

He stood on the seat of the cockpit and swung at them with the gun butt as snapping jaws tore at his leather jacket. First one, then another in quick succession would charge, while the remainder circled behind and attacked from the rear. He fought panic as he swung wildly in all directions, fighting to keep his balance. The beasts clawed at the back of his jacket and snapped at his hands. In desperation, he fired point-blank in all directions, killing and maiming as they attacked in mass. When the gun chamber clicked empty, he made a final gesture of defiance and hammered the gun butt into the skull of one of the beasts as they dragged him to the ground.

Carter opened the cockpit window and signaled the men to proceed on course. There was nothing to be done; it was time to go on. They had to find a clearing somewhere downstream where they could land and find shelter for the night.

Bernie peeked through the window by his seat for one last look at the scene below. His morbid curiousity of months past had given way to a deep sense of compassion for their lost comrade.

The wolves had reaped their reward and fought viciously for their share of the torn pieces of warm flesh.

The dismembered carcass left crimson trails in the snow as each beast dragged his chunk of meat across the field and disappeared into the shadows.

Bernie took off his glasses and wiped the cold sweat from his face. He looked across the aisle and watched Harry Ross shuffle a deck of cards for another game of solitaire. Bernie lowered his head to his knees and vomited on the floor.

Digger's make-believe heroics were far surpassed by the real life drama that he enacted today. The final scene was true to Digger's style, a hero to the end!

And for his encore, he willfully flouted the unwritten code of northland survival — "Save the last bullet for yourself."

A Speed Scout Wireless. Bob Tyson owned one of these airplanes. He was an avid race pilot and flew in many air races. It was built without the usual wire rigging between the wings. It was watercooled and the radiator was hidden behind the large spinner. Air for the radiator entered through a hole in the front of the spinner. (George Hardie collection)

14 HAPPY CHRISTMAS

Herol cast admiring glances at his reflection in Skrogg's Clothing Store window. He tried to appear nonchalant as he waited for Tyson who was inside bargaining with the owner for a bigger discount on the boxes of apparel he'd purchased for them.

The natty gray sport suit, white shirt and red tie, and black and white shoes gave Herol the air of a young playboy. All of this finery gave him a heady feeling as he looked at his reflection again. His eyes followed the shadows of late shoppers in the window's reflection, and he wondered if they noticed him. This was his first new store-bought suit and he hoped his secret was not obvious.

By contrast, Tyson wore a dark blue pin-striped suit, white shirt, striped tie, and black shoes. He wore his clothes with the casual air of an executive. He was truly handsome.

Snow flurries glided through the warm sunshine and piled up in tiny drifts across the sidewalk as Tyson and Herol strode down the avenue to their hotel. Their arms were laden with boxes and packages, the wealth of new clothing that would see them through the holiday.

Christmas was ten days away, and Herol still lingered in Saskatoon. He gave little thought to Fargo — that poverty-ridden existence held little attraction compared to the past several months of excitement. Best of all, he'd found that pot of gold Tyson had promised if Lady Luck smiled on them.

"You're a real sport in those duds, kid." Tyson grinned as he shifted the packages in his arms.

"You look better than Happy Bottom! I kinda miss my boots and breeches. How come we're gettin' all duded up?" Herol asked.

"Hell, we got the New Year's party comin' up that Happy Bottom is throwin' for everybody in the bank, and we're invited. Besides, me and Marge are gonna make a special announcement."

"I ain't ever been to a party. Do we have to dance and smooch with girls and all that stuff?"

"Guess so, kid. You don't have to smooch the dames unless you want to. Maybe Marge will give you a big smooch."

"I don't mind if Marge gives me a smooch. What are you gonna announce at the party?"

"I'll give you all of last summer's lumps if you tell anybody — we're gonna get married! Wanna be best man?" Tyson's eyes were beaming with happiness.

"Swell! How can I be the best man? You got the girl!"

Tyson chuckled, "Never mind. Me and Marge will explain it to you when the time comes."

They turned into the entrance of the plush Beverly Arms Hotel, their home since that last day two weeks ago when eight weary men and four equally tired aircraft landed in Saskatoon.

Early the next morning, Herol went out to the airport to help put skis on the Eaglerock.

Herol tugged on the bungee cord as he helped Pop Geraud stretch it taut from the wooden ski to the fitting on the fuselage. The Eaglerock would soon be ready for subzero winter flying and the long flight to Bemidji. The fuel and oil lines, oil tank, and crankcase were lagged with felt insulation and coated with waterglass. The twin core radiator was half covered with aluminum, and the front cockpit was cowled over to enclose an auxiliary fuel tank and a small storage area.

Tomorrow, Herol would test hop the bird and check out the riggin' on the skis before his trek south.

"I hear tell you and Kiddie Karr had your share of putting on skis," Pop said as he anchored a cotter key in the ski fitting.

"Sure did! Be a relief not to be thinking about pulling 'em off for wheels again."

"You want to check the skis out this evening?" Pop questioned.

"Nah, I'm supposed to have dinner with Bob and Marge. I gotta get back to the hotel and dude up in that new suit. I'll be out tomorrow about daybreak. If the bird checks out okay, maybe I can get as far as Winnipeg by afternoon."

"You gonna make it to Bemidji for Christmas?"

"Sure hope so. Want to see my granddad. Haven't heard from him for a while."

"You gonna be back for the New Year's shindig? Happy Bottom's throwin' a big bust for us. There's gonna be a party at the bank and a celebration here at the hangar on New Year's Day."

"Yeah, I'll be back in time. Just so I see Gramps over Christmas. What's with the party in the hangar?"

"You and Bob is expected to put on a show for the bank employees. They never saw some of the stuff they're hearin' about. You fellas got a big bonus comin' for this I hear."

"Cheez! Does Bob know about it?" Herol had visions of frostbitten hands clutching at flying wires.

"It was Bob's idea," Pop commented.

"Oh boy! One last chance to get creamed. I thought the

season was over. Guess I gotta make it to Bemidji and back before I worry about anything else," Herol laughed.

"Why sure, lots can happen by then."

"See you in the morning, Pop."

His departure the following day had been a routine affair. He'd said his goodbyes to his friends and made his way back out to the airport. Bob had given him a stern admonition to be careful and watch what he was doing and not to pull any dang fool kid's stuff.

At the airport Pop had the Eaglerock all warmed up and ready to go. Herol climbed in and with a wave to Pop, gunned the airplane around into the wind and poured the coal to her.

He banked the big biplane around after lifting off and came roaring back down the runway at ten feet above the snow. He pulled up in a power zoom over the hangar and waved one last time to Pop. Then he was gone into the brilliant blue of the winter sky.

Pop lingered outside in the frigid air watching the Eaglerock until she was just a speck on the southeastern horizon.

Herol slouched low in the cockpit. His goggles touched the inside of the windshield as he tried to escape the frigid blast that pounded his shoulders. The Eaglerock wallowed in turbulent air as he stomped his boots on the floorboards to combat the numbness in his toes that was spreading up his legs. He held the stick between his knees at intervals and sat on his mittened hands for the meager warmth his body provided.

Blowing snow reduced the world around him to an indistinct blur. The once familiar lakes appeared as snow-covered meadows, and the forests of evergreen were nearly hidden as snow blanketed the lush foilage and bent the branches to the ground. The only distinct landmarks were the logging camps or the tiny settlements nearby.

His southern trek had carried him over familiar territory.

From Winnipeg he'd flown the airmail route to Pembina, North Dakota, then to Grand Forks. The last leg was eastbound along a snowpacked road to Bemidji.

He'd been in the air nearly two hours since he departed Pembina, and when Grand Forks passed under his wing, he wheeled to the left and picked up the snowy ribbon of road to Bemidji. A strong quartering tail wind pushed him toward home but at the same time whipped the snow to new frenzies and reduced the visibility to near zero. He moved his frozen feet from the floor boards and pressed them against the rudder pedals. He pulled his hand from inside his jacket and retarded the throttle and sank through the blizzard until the Eaglerock's skis were nearly brushing the telephone poles along the road. He resisted the impulse to descend further, fearful of becoming snow-blind and losing all reference to the ground.

The cold was becoming more bearable as the numbness crept further up his legs. His hands were devoid of feeling and frozen into loose fists. He fought the drowsy feeling that overwhelmed him and moved his head from the protecting windshield to expose his face to the stinging snow that whipped past the cockpit. He felt revived for the moment, but knew that he would freeze to death if he didn't land soon. He had to stay awake.

A tiny settlement loomed ahead. He raced down its main street and quickly disappeared as he sped onward. He recognized the town but was to sleepy to recall its name. He had to keep that string of telephone lines in view and he had to stay awake for another half hour.

Another misery plagued him as he bounced his legs together to relieve the sense of urgency as his bladder threatened to burst. The thought of warm urine running down his breeches was sheer luxury, but he resisted the impulse, mindful of a more practical use for it when he landed. He squeezed his legs together again, forgetting his drowsy stupor. The bladder now occupied his attention.

"BAM!" The Eaglerock lurched violently to the right as wooden ski and wooden telephone pole slammed together. He climbed and turned left to avoid colliding with the next pole and prayed that he'd find his dangerous lifeline to Bemidji again. He peered over the right side of the cockpit and assured himself that the rigging was intact and the ski was riding normal. He might have a cracked ski but he'd worry about that if he was lucky enough to make a landing.

"Bob would sure chew me out if I got creamed this close to home," he mused. Then he recalled that time, eons ago, when Muggy Pederson took off from the meadow and almost made it home too.

"Ah, to heck with it!" he said out loud to wake himself up.

A pale glow crept over the ground ahead when he neared the outskirts of town. A truck crawled over the road, and a lone farmhouse winked at him with yellow kerosene lamps. The winter night would soon descend and threaten his chances for an already questionable landing. He sped over clusters of roof tops scattered beside the road and startled a team of horses that pulled a bobsled and protesting driver into a ditch. He caught sight of a shaking fist from the distraught driver as he passed overhead, then strained his eyes ahead to catch a familiar landmark. He held a straight course, he knew town was just ahead. The snow fury gave way to flickering street lamps and lighted storefronts as he roared down Beltrami Avenue and into darkness as he swung over Lake Bemidji. He stood the Eaglerock on her wingtip and wheeled quickly about to recover his landmark then climbed violently as a row of dim lights above his left wing threatened to collide with him. He sucked in his breath when he realized that he'd flown below the level of the bridge that spanned the Mississippi River.

A low turn over town gave him his bearings once again. He waved at the startled pedestrians, then settled on a southeasterly heading to his grandfather's cabin.

The wind abated as evening stars cast their pale light across the lake outlining the shores of Lake Plantaganet. He followed the shoreline at treetop level until he recognized the shallow cove that sheltered the old man's cabin, then turned toward the opposite shore. As the dark shadow of pine forest loomed ahead, he brushed through the top branches and turned into the wind. The engine was throttled back to a fast idle as he felt his way down through darkness to the icebound lake. The heels of the skis clattered against their restraining cables as they skipped across the soft drifts. He retarded the throttle to idle and the Eaglerock sank into deep snow. He guessed that Gramps' cabin was in darkness to the right of the lighted window ahead — that should be where old Spider lived.

The Hisso's propeller churned up another blizzard as the skis chopped through the drifting snow. He taxied cautiously as he neared the shoreline ahead, and when he approached Spider's lighted window he ruddered the Eaglerock to the right alongside the shore and stopped near the familiar path that led to Gramp's door.

He kept the engine idling to let it cool down, unbuckled his seat belt, and hoisted himself halfway out of the cockpit to shake the weariness that threatened to overtake him. The icy propeller blast revived him enough to help him clamber out of the cockpit. He shut off the ignition as he slid down the wingwalk and crumpled into the snow.

His mittened fists touched the side of his thighs and shot stinging pains through his legs. He reached toward his boots and felt nothing below his knees and realized that he couldn't walk. He elbowed across the snow on his belly until he reached the tail skid and packed a mound of snow around it. He pulled himself to a kneeling position and urinated to freeze the tail in place. His bladder still tormented him for further relief, but he held off until he shuffled under the wings and banked the heel of each ski with snow and wet them down. He sighed as his distended bladder was quiet once more. A giddy sense of well being enshrouded him and he fell on his face in the snow.

The room was barren except for a crude washstand, chair, and bed. A stovepipe protruded through the wall and exited into a nearby corner, affording only enough heat to combat the chill in the air. A window on the far side of the room was boarded over from the outside and a faded blanket hung halfway across the glass pane.

Herol was drowsy and vaguely aware of his surroundings. He savored the warmth of a feather bed and soft pillow under his head. His entire being tingled as new warmth coursed through his body. He didn't know how he got there, but he didn't care. It was great to be warm. Then a sense of uneasiness crept over him. The unfamiliar surroundings . . . something was wrong. Where was Gramps?

"Where in the hell am I? I gotta get out of here!" He threw off the heavy blankets and made for his boots that stood beside the chair. He collapsed to his knees on the first step.

The heavy timbered door creaked open and Spider stooped down and offered his rough hand to help the youth to his feet.

"That's pretty strong language you use since last time I saw you, Herol," Spider chided him.

"Guess so — Hi Spider." He took a faltering step and sat on the edge of the bed.

"You sit there a spell and I'll let you rest while I get some grub. I got some fresh Johnny Cake and honey. It's still hot from the oven. I made it for Christmas."

"Thanks. How'd I get here, Spider?"

"I saw you come in last night and didn't see no sign of you after you shut down your plane. I snowshoed out there and found you freezin' to death so I came back to

Any flying circus's reputation and livelihood depended on showmanship and ballyhoo. Imagination bolstered with raw courage was a prime airshow ingredient.

the cabin and got a dogsled to put you on and pulled you back here. You remember me walkin' you back and forth all night and all that hot coffee to keep you awake?''

"I don't remember much after I landed. You're a good friend Spider. Sure wish I knew how to say thanks. I'm grateful, I really am. Is Gramps okay?''

"I'll get you that Johnny Cake and coffee.'' Spider gave Herol a bleak look and turned his back and went to the kitchen.

"Christmas is only an hour away,'' he called from the other room. "Happy Christmas, Herol.''

"Happy Christmas to you,'' he shouted with a forced cheerfulness. "What happened to Gramps?''

The old man stepped through the bedroom doorway, his arms laden with a brown crock full of Johnny Cake, a coffee pot, and cups. He pushed the chair across the room with his foot and set the food down. Then he sat on the bed next to Herol.

"Me and Bo was good friends for over fifty years. I miss the light in his window and that makes me more lonesome 'cause I know he ain't there.'' The old man's eyes were closed and his fists were clenched together as he silently pounded them against his knees.

"He's dead, son. Day before yesterday. He told me he was plumb tuckered out from waiting, and he didn't know what he was waiting for anymore. He was waiting for you, but he didn't believe you were alive. He always felt good when you sent him those pictures from all those places you was at. Between postcards, he figured you got killed until the next card. Maybe the worry got him tuckered out and he couldn't hold on 'til Christmas.''

Herol said nothing. He couldn't. One of Bob's sage remarks flashed across his mind . . . "Kid, you know too much about dying and nothing about living . . .''

This was different. This was Gramps! He couldn't forget Gramps like Bob had taught him to forget all the other 'Johnnies'.

"You all right son?" Spider touched his hand.

"Yeah, I'm okay. I know a lot about dying. I die a little each day, and so do the other guys. I should have died instead of Gramps."

"Now son, I'm sure Bo understood."

"Why didn't you let me freeze to death the other night? I could have spent Christmas with him." Herol rested his head on the old man's shoulder and stared at the floor.

"Happy Christmas, Gramps," he muttered as tears slowly made their way down his cheeks.

Bob Tyson's aerial extravaganzas were seldom the same. The 'Great Diablo's' feats were always 'played by ear' and always a surprise. He was a natural showman and a giant among his peers. (Bill Rhode collection)

15 HAPPY BIRTHDAY

Blue-white diamonds glittered in the path of the winter sun as it rolled across its arctic zenith, casting long shadows beside green pines heavy laden with snow. The Canadian wilderness was a snowscape of desolate splendor, seemingly barren of life until the forest creatures emerged in the spring.

The solitude was shattered by a red and silver biplane roaring through the still air at ground level. It hedge-hopped over a vast patchwork of scrub timber and open fields and disappeared beyond the horizon.

"Hot damn! This is the life!" Herol yelled.

He watched the stunted trees whiz past his wing tip. As the field ahead spread into a rolling white prairie, he touched the fresh snow with his skis and listened to their clatter as he high-tailed along, racing the whirling snowstorm that churned through the propeller. At the far end of the field, he lifted the skis clear of the snow in time to miss a thicket of tall pines that rushed at him and flew into another blizzard as he threshed his way through the snow-laden branches and wallowed out of the top.

"Wow! That was close! Better straighten up for a while! Bob would be mad if I busted up before New Year's Day." Herol laughed at his own grim sense of humor.

He climbed in a lazy circle to about three hundred feet, then resumed a northwesterly course. He marveled at how serene the wilderness appeared today compared to the howling nightmare he'd fought just a week ago. He craned his neck in all directions inhaling the clean blue around him. He knew for certain that he could see a mile beyond eternity. Life was good!

He clung to his brief moments of delight, knowing they were his greatest reward for months of fear and broken bones. He felt the need to live each moment to the fullest and cram a hundred years into a short life span. He dreaded the New Year's show with a morbid fear he'd never known before. There were few precious days left.

"The only dumb thing Bob ever did was to agree on that show in the middle of winter. By damn! If I slip and bust my fanny I'm gonna be laughin' when I hit the ground 'cause I've lived my dozen lifetimes and I ain't lost a thing!" Herol examined his strange brand of philosophy and realized how akin he and his partner had grown.

He flew onward across the vast stretch of prairie until the cropland was devoured by encroaching timber that spanned the province. His hellbent-for-leather mood had given way to more somber thoughts as the last bit of prairie slid under his wing and dense forest and rocky outcroppings appeared at increasing intervals. He sensed that a forced landing would be fatal and stole a quick glance at his engine instruments for assurance. A fleeting thought of Digger's viscious exit crossed his mind, but he dispelled it quickly knowing that he would save a .45 slug for himself.

He reached down and felt the bulges in the sides of his boots and was assured that he hadn't forgotten the extra clips of ammunition. Then he fingered the butt of the .45 automatic under his jacket and knew it was loaded.

"Why worry about a forced landing. I almost busted my fanny at Christmas and its damned near a gut cinch for New Year's." The strange dread that he felt before came over him again.

"So what! Live it up, die young and have a good lookin' corpse!" He thought the morbid joke would lift his spirits, but it didn't.

Before he'd left Bemidji, Spider had insisted that he take the old man's trapping clothes along with him. Old Spider'd even stood right there and made him put them on over his clothing before he got in the airplane. Now he was happily surprised to discover that he was nice and warm even though the temperature in the cockpit was subzero.

The buckskin moccasins and leggings that encased his boots gave off a rancid odor from numerous applications of bear grease. The fringed buckskin parka that hung loosely over his jacket emitted an equally pungent smell.

The strips of beef jerky and smoked venison that Spider stowed in the front cockpit would provide nourishment for about two weeks and there was even a piece of Johnny Cake left over from Christmas.

Herol wished that he had recognized the old man's wisdom when he accepted these gifts and decided that if he survived New Year's Day, Spider would receive the picture postcards and visits the same as Gramps had.

For three days he'd flown northwest. Now finally his lonely odyssey was nearly over. The harsh terrain was far behind, and beneath him level fields crisscrossed with country roads gave an appearance of safety. The feisty Eaglerock barked her throaty roar while Herol slouched comfortably in the cockpit intent on spotting a familiar landmark that would lead them to Saskatoon.

The northern latitudes provided meager daylight this time of year as the winter sun remained low in its orbit across the horizon. It was a little past noon and already shadows were reaching long fingers across the snow. The clear blue dome overhead was tinted with pale yellow. Far to the northwest, brush strokes of red-orange were painted across the sky foretelling stormy days ahead.

A moving column of white smoke emerged from the backdrop of low hanging clouds in the distance. It was probably one of the long freight trains heading south, laden with pulp wood destined for the paper mills.

The Eaglerock lazed into a gentle left turn to intercept the train several miles ahead and as the silver wings flashed against the sun, an array of steel tracks on the ground mirrored their image from the slanting rays of light. There were several pairs of rails, so it was probably the busy roadbed between Saskatoon and Moose Jaw.

Herol leaned into a wide arc, descending in a shallow dive as he drew abreast of the speeding locomotive. The white smoke rings that puffed from the engine's stack merged in a steady white streamer. Then he heard the friendly "whooooo" from its whistle. Herol inched up alongside the engineer's cab and waved. They raced neck and neck down the track while each crewman exchanged greetings with him. Finally, he waggled his wings in farewell and disappeared over the top of the cab. The Eaglerock zoomed into a climbing turn and headed in the opposite direction toward home.

Two black columns of smoke on the distant horizon rose vertically in the calm air and dissipated into a dark haze as they touched the base of the overcast sky. This landmark signaled the end of a hazardous adventure for the boy and his plane. It was the busy rail terminal a few miles from home.

A light chop developed in the calm air as he approached the grey overcast that was spreading over the city. Patches of low strata drifted beneath the moisture-laden clouds that hung in the north. A winter storm was brewing and he smiled at the pleasant thought of a New Year's cancellation.

The snowbound runway passed under his left wing. He banked into the wind and hummed a tune as he glided in for a landing. He'd made it!

Tyson yawned as he roused himself from a short nap and got up from his chair. He took a quick step across the room and turned his back to the potbellied stove.

"The kid's home," he commented dryly.

"How you know?" Pop asked as he looked out the hangar office window.

"Heard a Hisso fly over, then it died out. Figured he flew downwind and turned in for a landing."

Pop hurried across the hangar and peeked out the front door. "He's right out front, engine is still running. He must have just taxied in."

"Yeah, I know," Tyson said. "Lemme talk to him before anybody else gets to him."

"I understand. Call when you need me. I'll drive you into town."

"Thanks."

Herol shut down the engine when he saw the hangar door open and Tyson stepping out. He unbuckled his seat belt and instinctively reached for his parachute harness, then realized he hadn't worn a chute in over a month. He rubbed his shoulders and thighs and discovered that his harness burns were all healed.

"You scratchin' for fleas, kid? Glad to see you." Tyson gave him a gentle nudge and helped him out of the cockpit.

"Hiya Bob! Nah, I ain't got no more harness burns. Glad to see ya." Herol grasped his partner's hand.

"Proud of you kid. Nobody thought you'd get back alive except me. Had to stick up for my partner, right? And when Happy Bottom found out that Bemidji was clean out of the country instead of some little jerkwater town around here, he scratched his rear for a week! And he's holding two thousand bucks for the Eaglerock in case you didn't make it. So, you got an airplane if you want it — in case you didn't hear me the first time, I'm proud of you, kid!"

"Gee thanks! And I'm glad we're partners. You're darned right I want the Eaglerock!"

"Got something to talk about — something that happened while you were gone," Bob said.

"Me too," Herol interrupted. "Wanna tell you all about the trip." Then his face clouded over as he thought about Gramps.

"Let's get you inside to warm up a bit then we'll go to town. Lots to talk about."

"What's the matter Bob, you ain't feelin' good?"

Herol looked at his friend's ashen features and saw eyes that bespoke rage and anguish at the same time. He'd grown to know this strange man and respect him and wished that he understood.

"Just a headache, but it's okay."

"Hello, son," Pop greeted him as they sidled up to the stove.

"Hi, Pop. Got the bird back in good shape, and thanks to you for all your help."

As Herol relaxed by the side of the stove, old Spider's buckskins came to life and before long he was smelling like the Grand Champion of the Skunk Works!

"Goddamnmit! Take that Pocahontas suit off! You smell worse than a fresh boogered bear!" Tyson growled and headed for fresh air in the hangar.

"You stink!" was all that Pop could say and went out the door.

Herol examined his rustic apparel, gave it a sniff and agreed. "Guess I do. It didn't smell this bad in the plane."

"Take it off and stop talkin' to yourself," Tyson hollered.

"Yeah, right away. I got this stuff from a friend. It's good and warm."

"Yeah, and it stinks, too!"

The Model A labored down the road, Pop feeling his way through tracks hidden with fresh snow. A broken chain on a rear tire banged against the fender with monotonous regularity and measured their progress to town. Snowflakes sprinkled the warm hood of the truck and melted, then trickled down the sides and froze in tiny rivulets.

They huddled together on the narrow seat — Pop behind the wheel, Herol in the middle, and Tyson squeezed in the corner trying to find room for his legs.

"Hey, maybe the New Year's shindig will be cancelled," Herol said hopefully.

"Don't count on it," Tyson grunted.

"Carter is gonna test-hop the Standard. I overhauled the engine and put a new prop and skis on. We got a buyer in the States. Wanna use it for the show?" Pop asked.

"Might. Let you know tomorrow," Tyson answered.

"How was the trip?" Pop hoped to keep the youth talking.

"Swell! Had a warm place to sleep every night and lots of free grub. And that extra gas tank sure saved a lot of worry."

"Where did you get gas?" Tyson asked.

"I'd hunt up a farmyard just before dark and land. The farmers usually had tractor gas and they'd sell me what I needed. Sometimes I'd get supper, and I could always sleep in their barns."

"Guess you made out like a bandit," Tyson commented.

"One guy that lived near the border wasn't very friendly. He didn't sell me any gas, so I walked about three miles to a station and they delivered some. The farmer got to talking to the driver and from what I overheard, he figured that I stole the plane and was gonna make a bootleg sale across the border. I think he wanted the driver to get in touch with the law when he got back to town."

"Did you get pinched?" Pop asked.

"Nah, I guess he didn't call the law. Anyway, the old

farmer went to bed early and I slept in a haystack a few hours. About two in the morning, I decided to hightail it out of there just in case the sheriff came around after daybreak. I hauled out and flew the rest of the night. It was spooky as hell not knowing where I was or what was underneath me. I didn't do that again until the snowstorm near Bemidji and had to land at night."

"Pretty good," Tyson smiled.

"Any mishaps?" Pop asked.

"Damned near busted the right ski — bounced it off a telephone pole in the snowstorm! I hope it's okay."

"We'll check it."

"Did Bernie get the job Happy Bottom promised?" Herol asked.

"Uh, that's what I want to talk to you about," Tyson said.

"Hope you didn't fix me up with an office job that Bernie didn't want! I'm gonna keep flyin'!"

"Nope, Bernie couldn't take the job. Something happened — he's just another 'Johnnie' now. He got creamed three days ago."

"Bernie! You're kiddin'! He ain't been flyin'! How could he get creamed?"

"It's true," Tyson spoke softly.

"He was my buddy!" His hands shook, then his entire body trembled. Maybe he was exhausted, or maybe it was the gut feeling he'd lived with for days.

"Cheez," he wiped the sweat from his forehead. "My grandfather died just before I got home, and now Bernie. What in hell is happening? And don't tell me that Gramps is just another damn Johnnie like Bernie!" He lashed out at Tyson.

"It's alright, kid," Tyson laid a comforting arm around him.

"It happened quick — he didn't know nuthin'," Pop said, and brought the truck to a stop in front of the Beverly Arms.

"C'mon, let's get cleaned up and grab us a steak." Tyson slammed the truck door shut and led his partner across the walk to the hotel.

Their snug apartment was a far cry from the many nights when Herol slept in a farmer's barn and shared his bed with a herd of bawling cows. This was living and he loved every minute of it.

He soaked in the bathtub that was near overflowing with soapy water, intent on scrubbing a week's grime out of his skin. He sniffed his lathered shoulders and wondered if Bob would smell the bear grease after he'd rinsed the perfumed soap off. He thought of Bernie and regretted that he didn't have a chance to say goodbye before he left. It was hard to believe that he was dead — and just another 'Johnnie.' He wondered if he could forget his friend.

"The dining room closes around eight o'clock," Tyson hollered toward the bathroom door.

"Out in a minute. Let me know if you smell that bear grease," Herol shouted through the door.

"You'll smell like a buttercup after the time you've been in there!"

"Could you tell me what happened to Bernie? I can't believe it!" He wrapped a towel around his middle and dripped his way to the bedroom, intent on fingering soap from his ears.

Tyson got up from his chair and followed him, then stopped and leaned against the doorframe. "You look cleaner than a she mouse's belly!" he tried for openers.

"Feel good! And no more lumps that need to heal up, either," Herol grinned. "How did Bernie get creamed if he wasn't flyin'?" he blurted.

"Remember, Pop said he overhauled the engine and put on a new prop on the Standard? Well, he had to because of that damned squirrel's clowning around while Carter was running the bird. He was dancing around like a hot-pants cuckoo and he backed into the propeller. It split him right down the middle, from between his eye-glasses to his belly button. That damn 'Johnnie' is better off, wherever he is!" Tyson's eyes were pools of ice water and his features like granite.

Herol stood rigid, not believing the brutal tirade he'd just heard. He knew this man was strange, but he thought he understood him — even loved and respected him. His first impression was right after all. He was a demon!

"You sick animal! You don't deserve to live! Why didn't you get killed instead of Bernie?" Herol thrashed around the bedroom and made his way toward Tyson who remained motionless in the doorway. He threw a wild punch at him and missed, then swung again and again.

"Easy now, just wanted to give it to you short and sweet," Tyson spoke in a flat tone.

Herol crouched in front of him, fists doubled, undecided whether to throw another punch or not.

"I hope your guts are splattered all over that field on New Year's Day instead of mine. I'm gonna do my own thinking for a change, too. I don't give a damn what happens to you!"

He backed away and slumped across the arm of the sofa behind him. He started to cry and felt angry because Bob would be mocking him. When his anger subsided, he sat up and stared at the wall seeing nothing.

Tyson moved slowly across the room, stood in front of the youth for a minute, then sat down beside him.

"I didn't mean none of that raunchy stuff about Bernie. I like him too. Just needed to shock hell out of you and get you pissed off so you'd get rid of them feelings inside you that was building up and maybe explode at the wrong time. I'm glad you took it out on me." Tyson was very gentle.

Herol turned toward him and leaned on his shoulder. "Gee! Wish I'd grow up," he sobbed.

"You're a hundred years old today," Tyson replied.

"Then why do you always call me kid?"

"Sorry. Just habit I guess."

"Wish I hadn't said all them things to you. I don't want you to die. Honest!" He was on the verge of tears again.

"Cut out the crybaby stuff. That's for girls and old ladies at funerals. You're a hundred years old, remember? Happy Birthday!"

"You're right. I hit a hundred today. Time to forget the 'Johnnies'."

"You're back on track now," Tyson smiled, knowing that all was well.

They ate supper in silence, each one immersed in his own thoughts.

Tyson fretted about the coming show, then turned his thoughts to Marge to dispel his misgivings.

Herol tried to forget about the New Year's show. His thoughts drifted to Gramps and Bernie, and all the others — Muggy, Dick, Johnnie, Digger, Black Jack and Dutch. He reminded himself of what Bob always said, "Forget the 'Johnnies', they're gone."

He decided to make an exception to the rule and remember Gramps, he wasn't a 'Johnnie.'

Marge Savageau and her little sister Rosella, taken at Saskatoon, Saskatchewan, Canada. Marge and Bob Tyson were supposed to be married on New Year's Day.
(Author's collection)

16 HAPPY NEW YEAR

HB was in an amiable mood as he mingled with the crowd of employees that filled the bank lobby. He was a gracious host and prided himself in being able to recall the first names of everyone present. He made them feel very special, and indeed they were. He was fond of his faithful employees and they in turn responded with warmth and goodwill. He made a mental note to compliment Marge and her staff of helpers tonight for the many hours they'd spent cooking and baking the luncheon snacks for this New Year's Eve party. Their holiday decorations throughout the bank also added a personal warmth to an otherwise formal setting.

The walls of the lobby were adorned with silver tinsel, holly wreaths, and twinkling lights. Even the exquisite marble statues that stood in strategic areas were bedecked in holiday fashion — Queen Victoria, Augustus Caesar, Lord Nelson, King Edward, and the rest were crowned with a sprig of mistletoe. No doubt HB took a dim view of this bit of whimsy, but he said nothing.

A huge evergreen commanded the center of the floor. Soft colored lights winked from behind the lush foilage, and tinsel-laden boughs cast a multicolored hue on the mountain of gifts piled under the tree.

A thin net, bulging with hundreds of balloons, was

suspended from the ceiling awaiting release at midnight.

Tables were aligned against the wall and groaned in protest under the weight of party snacks and bottles of holiday cheer. And at the end of the line, Pop Geraud and Carter rubbed elbows with each other as they worked at top speed mixing drinks for the thirsty crowd. In a nearby corner, an orchestra blared out noisy renditions of dancing favorites.

It was nearly midnight before Marge and Bob decided to announce their forthcoming wedding plans, adding a new dimension to the coming year, only minutes away.

Tyson spoke a few words to the orchestra leader, who ended the music, then, after a brief pause, began anew with a loud fanfare.

For the first time in his life, Tyson was unsure of what he should do next. He gave the expectant crowd an embarrassed grin and headed toward Marge and Herol who stood nearby.

"You ain't gonna say nuthin' about that best man stuff, are you?" Herol asked when he arrived.

"Well, I dunno — something like that," he replied.

"Say something, Bob — they're waiting," Marge prompted.

"Uh, I've never done this before," he fumbled.

"How about you, Herol?" Marge smiled.

"Bob always tells me to keep my mouth shut!"

Friendly titters from the audience only heightened the nervous anxiety of the partners.

"My two heroes are frightened! I'll have to make the announcement, I guess," Marge teased.

She clasped Bob and Herol's hands and raised them above her head for a moment, and when the crowd was silenced she revealed the well kept secret of their wedding plans. Amidst a burst of applause, she invited everyone to attend St. Phillips Cathedral tomorrow evening at five o'clock for the wedding ceremony.

As if on cue, an avalanche of colored balloons with

confetti streamers cascaded from the vaulted ceiling and showered the crowd.

"HAPPY NEW YEAR!" everyone shouted.

The good wishes, handshakes, and happy bedlam seemed to last forever, and probably the most surprised of everyone was HB, who grew impatient waiting for the endless round of congratulations to subside. As usual, he rubbed his bald head to a rosy pink while he waited for a degree of silence. Finally he raised his arms for attention and waddled toward the happy couple. He shook their hands and offered warm congratulations, then bestowed a hesitant kiss on Marge's cheek. He abruptly turned his back on them and faced the audience to announce that Tyson would be presented an additional bonus tomorrow after the airshow. And since this impromptu show was not too many hours away, Tyson would be expected to provide a safe and sane exhibition tomorrow, and not the suicidal performances he was known for. As an afterthought, HB pointed a chubby finger at Herol who stood nearby.

"That means you, too!"

Daybreak was yet an hour away when Herol tiptoed out of the apartment, leaving Tyson content with his dreams. He rode the elevator down to the lobby, glanced wistfully at the "CLOSED" sign on the lunchroom window, and walked through the front door and out into the frigid gale wind.

He pressed the fur-lined parka against his ears, stood with his back to the wind, and stomped his feet in the drifting snow. He listened for the sound of Pop's Model A truck. His sleepy eyes watched a line of street lamps disappear whenever gusty winds lashed the fallen snow, then reappear wearing a pale halo of ice crystals.

He was pacing back and forth on the sidewalk when two yellow lights emerged through the blowing snow, and the familiar chug-chug of Pop's truck broke the early morning silence. He stepped to the street and reached for the door that swung open as the truck's wheels crunched to a stop in the snow.

"Mornin' son! — Jump in — ain't much heat in here," Pop shouted.

Herol settled himself onto the cold leather seat and pounded his boots against the floorboards to encourage a bit of warmth into his numbed feet.

Pop reached into a pocket of his sheepskin coat and withdrew a pint bottle of moonshine.

"Have a snort! It'll warm you up!"

"Thanks, I don't drink much booze. Bob would flatten my head if he caught me."

"It's okay this time — medicinal purposes. You're cold, ain't you?" Pop shoved the bottle in front of Herol.

"Yeah, thanks." He upended the bottle and gulped down a slug of fiery brew, then gasped for breath and wiped the tears from his eyes.

"Wow! That's wild!" he whispered, then smiled as the warm glow crept through his body.

"I won't tell Tyson if you don't," Pop volunteered.

"Yeah, we better not. That stuff's awful, but it feels good when it hits bottom!"

Pop retrieved the bottle, took another healthy swig, thumped the cork into the neck and put it back in his pocket. Then he focused his attention ahead on the indistinct road to the airport.

"I'll bet Happy Bottom cancels the show today," Herol ventured hopefully.

"Maybe, but I'm getting the Standard and the Eaglerock ready just in case the wind lets up. Best you check the ski rigging on both ships this morning. Tyson said something yesterday about one of you making a plane change on the ladder. I got it coiled and rigged on the heel of the left ski on the Standard. I think he plans to fly the Standard, and you and Carter will be in the Eaglerock. Sounds like you're elected to climb the ladder! — I think he's gonna make a jump, you land and pick him up. It's gonna be short and sweet. He's anxious to get married!"

"Wow! It's too cold to do anything today. He didn't say nuthin' last night about any routine — he musta been thinkin' about Marge. I bet he's sorry Happy Bottom talked him into this. I sure am — it's crazy doing this in the winter!"

"Maybe he wanted the extra money to get married," Pop suggested.

"Maybe. He wouldn't get suckered into something like this any other time."

"What you plan to do for the rest of the winter?" Pop asked.

"I plan to fly back to Bemidji in a couple weeks, after I drop Marge and Bob off in Ottawa. I'm gonna live in my grandfather's cabin. Old Spider lives next door and is pretty lonely; maybe I can help him."

"Don't suppose Tyson will hit the airshow trail next spring now that he's getting married?"

"Probably not. They're spending their honeymoon in Ottawa, and while we're there Bob is gonna talk to Jan Carver about getting a job for me in the spring. He's starting up some kind of flying for the weather bureau, and he's looking for good pilots with their own planes. That's me, I hope."

"I know Carver," Pop said. "He's from Owen Sound, but he's been flying in the States for the last few years. I hear tell there's lots of government experimenting in trying to find out how we can fly through the weather instead of underneath it."

"What's wrong with hedgehopping when the weather is bad?" Herol asked.

"It just ain't safe — too many church steeples and high towers to get tangled up with when you can't see. Come to think of it, you smacked a telephone pole near Bemidji this winter."

"Sure did! Even smacked a few tree tops on the way back, but I was horsing around!" Herol boasted.

Pop decided to ignore the youthful exuberance and changed the subject.

"You figure both of you will be working in Ottawa?"

"Bob says I'll probably be in Winnipeg, or across the border in Pembina, or Fargo, North Dakota. I hope it's Winnipeg 'cause he's gonna fly mail out of there."

"So Tyson's going to fly the mail out and you're going to be a weather pilot — sounds great!" Pop smiled.

"Guess I'm too young to be a mail pilot, and that weather stuff sounds screwy to me. Last fall we crossed the continent and got through the blizzards. Digger was the only one that didn't make it. I'd just as soon hit the airshow trail again next spring."

"Stick to what Tyson thinks is best for you. Don't press your luck. You've been living on borrowed time for a long while," Pop cautioned.

"Ah nuts!" Herol grumbled.

They rode in silence for several minutes, Pop's wisdom vying with Herol's youthful impatience.

"It's cold! Want another drink?" Pop broke the silence.

"You talked me into it!"

The wind had dropped by the time Pop and Herol arrived at the field. A column of grey-white smoke rose vertically from the chimney above Carter's office, a sure sign that coffee was brewing atop the potbellied stove.

They clambered out of the meager warmth of the truck and into the frigid morning air. Blue skies spanned the horizon and the arctic sun showered the fresh snow with diamonds. Their boots crunched into the powdery snow and clouds of vapor surrounded them as their breaths crystalized. They stepped into the hangar, stomping their feet in the doorway, then walked across the concrete floor to Carter's office.

The hangar was spotlessly clean. In place of the aircraft, two lines of tables supported coffee urns, stacks of paper plates and cups, and silverware. Dozens of folding chairs leaned against the table awaiting the arrival of last night's New Year's Eve party.

"Come and get it!" Carter called from his office.

"Coming!", they shouted, and made a beeline through the door toward the warm stove.

Carter handed them mugs of steaming brew, then poured one for himself. Pop reached into his sheepskin and uncorked his bottle of moonshine and poured a generous draft into Carter's cup, then one for himself, and offered the bottle to Herol.

"Thanks, better not — Bob would twist off my head and hand it to me if he caught us."

"Yeah, guess so," Pop smiled.

"That reminds me, in case you plan to wear them buckskins today, don't put 'em on in here. They get mighty ripe indoors."

"Yeah, I remember, but they sure get warm."

"And so does the bear, but he stinks too!" Pop said.

Carter lowered his head behind the windshield and waited patiently for the big Hisso engine in the Eaglerock to warm up. It would take several minutes for the oil to reach normal temperature in the subzero chill. He glanced out the cockpit past the right wingtip toward the Standard that was also being readied for the New Year's Day flight.

Pop was pressurizing a gasoline heater that he would place under the canvas tent enclosure around the engine, while Herol was busy checking cables, rubber bungee cords, and fittings on the ski rigging. The calm arctic air magnified the sounds of activity around the two aircraft.

"Wish I was wearin' them smelly buckskins of yours, I'm cold!" Pop grunted.

"Ha! I'm warm!" Herol laughed. "Why don't you crawl under the engine cover with the heater? I'll finish up."

"Say now, that's a pretty doggone good idea," Pop said. "I think I might just do that for a minute."

The muffled sound of several automobiles grew more distinct and soon a procession of black cars was visible crawling along the road. HB and company would soon arrive to attend the New Year's Day festivities.

The Eaglerock's engine ticked to a stop and Carter clambered out of the cockpit, stepped down from the wingwalk and trudged through the snow to retrieve the engine cover that lay nearby. He dragged it behind him until he reached the nose of the plane, then flung it over the warm cowling and pulled the straps tight around the bottom to retain the heat. He glanced at the melted snow beneath the engine that had refrozen to ice and anchored the skis to the ground.

"Herol! Don't forget to use a crowbar and pry these skis loose — they're stuck!"

"Yeah, okay. Soon as we get this bird warmed up," he shouted back.

"I'll take care of the skis," Pop said as he stuck his head out of the canvas enclosure beneath the Standard. He crawled out from his warm haven pulling the gasoline heater with him.

"Let's try to get this engine started." Pop folded the engine cover and pulled it and the heater out of the way.

"Wanna give 'her a twist? Switch is off!"

"Okay, whenever you're ready."

Pop hoisted himself into the rear cockpit while Herol swung the wooden propeller, sucking raw gas into the carburetor. He rocked the propeller over number one cylinder and stepped back.

"Hey Pop! See if it will start on the booster — all clear!"

Pop gave the booster coil a hefty twist, and the Hisso belched an oily cloud of smoke, and started.

"Hot dog!" yelled Herol.

Back in town, Tyson stood by the living room window and watched the frenzied snow outdoors. The street was being lashed by nervous winds gasping for breath as the blizzard faded. Patches of blue peeked from behind low hanging clouds that were rapidly dissipating. He ground his cigarette in the ash tray, glanced out the window again, and decided that Happy Bottom's airshow would come off as scheduled. He wondered when Herol left, knowing that he was already at the field. A faint smile crossed his features as he thought of the half-starved kid he'd first met in Winnipeg, and how he'd matured into the seasoned airman of today. The promise he'd made "Cash" Chambers that rainsoaked night in the bus depot had been fulfilled. The kid was able to take care of himself, and with a little help from Lady Luck, he might survive beyond a pilot's normal short life span.

Today would be the last flight for both of them, and though he'd performed hundreds of times before, something was out of kilter this morning. Perhaps it was the anxiety of his new responsibility to Marge, or it might be the farewell to his air circus world he'd known so long. Something was wrong, but damned if he could figure it out.

He dismissed his thoughts and rose from his chair and went to the bedroom to finish dressing. It would soon be time to meet HB and Marge at the bank, then to the field and showtime.

Having dressed and made it to the bank on time, Tyson slouched against the counter of a teller's window in the bank lobby. He watched the janitors with idle interest as they descended on the debris from last night's party. There were mounds of confetti, yards of discarded gift wrappings, and hundreds of deflated balloons everywhere. Two men were scaling ladders against the wall and about to retrieve the holiday decorations that would be stored for another year.

He watched Marge and a host of her helpers clean the

tables of the leftover feast, while others scurried about making themselves generally useful. A line of industrious sweepers abreast the opposite wall descended on him with their pushbrooms, so he sidestepped out of their way and took refuge in the vacant teller's office.

He found a week-old newspaper beneath the counter and tried to catch up on a bit of the local color. He was unable to concentrate. The distractions in the lobby made him nervous and he wished that they'd finish soon. He was anxious to be in his own element again.

He reminded himself to double-check Herol's harness and the rigging of the ladder on the ski of the Standard. He didn't want a mishap to mar this special day — his wedding. He felt a strong compulsion to get to the field. There were so many things he should tell Herol, yet he wasn't sure just what it was that needed saying. His thoughts were interrupted when Marge brushed his cheek with a kiss.

"So here you are!" she smiled. "Give me five minutes to freshen up and I'll meet you at HB's car." Another radiant smile and she was gone.

The trio crowded around the stove in Carter's office sipping hot coffee while they waited for HB's arrival.

"No more booze! Too close to flight time." Pop announced.

Carter nodded in agreement and Herol grinned.

There was a flurry of activity outside the hangar as car doors opened and slammed shut accompanied by shouts of "Happy New Year!"

"I guess they're here," Carter remarked and stepped through the office door into the hangar to greet his visitors.

The side door entrance near the front of the hangar opened and HB, Marge, Tyson, and a host of guests crowded through the doorway.

"HAPPY NEW YEAR!" they chorused.

Carter returned their greetings and Pop and Herol waved from the office door.

The happy couple walked arm in arm across the floor towards Carter's office and HB remained in the hangar with the others who began unfolding chairs and setting them in place while some set covered dishes of food and party snacks on the tables.

"Cheez! You guys are early!" Herol greeted them.

"Ain't you gonna say good morning to the future Mrs. Tyson?" Bob smiled.

"Uh — good morning," Herol blushed.

Marge stepped close to Herol, gave him an affectionate hug, then backed away in surprise.

"It's them smelly buckskins! Herol swears by 'em, and we swear at him!" Tyson grinned at her.

Marge stepped close to Herol again and hugged him, holding her nose with her fingers in mock disapproval.

"Since this is our last flight, I can stand the smell if the rest of you can!" Tyson grinned again.

"Sure! Okay! Why not!" They laughed.

"Hear tell you're gonna pair up with Gimpy down in Winnipeg," Carter said.

"Yeah, he's the operations manager and I'm gonna fly mail — this might be one of my stops."

"And no wingwalking during your stops!" Marge teased.

"I'll be able to make a good home for us." Tyson smiled at the girl.

"Sounds great, and good luck to both of you," Pop said.

"Goodies are on the table! Come and get it!" someone shouted from the hangar.

"Marge, I gotta talk with Herol about the flight. Why don't you join the others and I'll see you in a little while." Tyson smiled at her.

"All right, see you soon." She pressed his arm for a moment, then left.

Pop buttoned his sheepskin coat and pulled a woolen cap down over his ears.

"I'm gonna see how much the engines have cooled down. Might have to start them again soon."

"Call if you need me; I'm goin' to talk to HB," Carter said.

Tyson moved away from the stove and stepped over to the window. He stood there for several minutes, looking at the snowbound world outside. Finally he spoke.

"I sent a telegram for you to Jan Carver. He's expecting you in a week."

"How come? Sounds like I'm goin' alone," Herol asked.

"He don't know about me and Marge. Just in case we can't make it, I got your tracks covered. You'll be okay."

Tyson turned away from the window to face his partner. His manner was gentle and the harsh features that he wore so often were gone.

"Gee! I don't understand — we're still partners, ain't we? I don't want to leave alone."

"Sure, we'll always be partners." Tyson's voice was barely audible. "I'm saying just in case we can't. And remember, I still do the thinking for both of us."

Marge stopped short in the doorway. She immediately sensed the strange mood that prevailed between the two, and for a fleeting moment had misgivings that something was wrong. She forced a smile on her face and walked in.

"Aren't you boys ever coming out? Can I bring you a sandwich or something?" She stepped close to Tyson and laid a gentle hand on his arm.

"No thanks, Honey — was just going out to my locker to get a harness for Herol. Walk me out, okay?"

"Alright," she smiled, and they were gone.

"Somethin' ain't right. Sure wish I could figure him out," Herol mumbled.

It was time to get on with the task at hand. The hangar luncheon would soon be over and Happy Bottom was expecting the finale of the day's entertainment.

Tyson returned with Marge on one arm and a leather harness on the other.

"I brought my harness instead of yours — it's got a better saddle on the bottom. Pull off that Pocahontus suit and we'll get you rigged up."

"May I watch?" Marge asked.

"Promise not to give away our trade secrets?" Tyson smiled.

"I promise!"

Herol was glad that she was here. Maybe she shared his feelings. He wondered if they'd ever understand this strange man.

Tyson busied himself adjusting buckles on the leather straps while Herol pulled the buckskins up and over his head. He felt Tyson's hands lend a much needed pull and the pancho slid over his shoulders.

"Thanks, it doesn't come off easy!"

"You've got too damned many clothes on underneath! Strip down to everything except your long johns and the woolen shirt. You'll get creamed sure as hell if you don't!" Tyson was back to his normal harsh concern for his partner.

Herol blushed as he followed instructions and hoped that Marge wouldn't look at him in his long johns.

"I'll leave you two with your trade secrets for a few minutes." Marge tactfully left the room.

"Come back soon," Tyson said softly.

They busied themselves adjusting the leather straps to fit. Tyson was more attentive than usual. He realigned the canvas saddle that would support Herol's body and carefully adjusted the double straps that extended from the shoulders and hugged both sides of his arms to the wrist.

"Cinch them straps around your wrists and above your elbows tighter, then I'll shorten the hooks. They're a little too far over the center of your hands — you might not be able to get a sure grip on the ladder." Tyson seemed very gentle.

Pop stuck his head through the doorway.

"You guys about ready? We oughta get them engines running soon."

"Get out of here!" Tyson flared up.

"Sorry."

"You feel okay? Wanna cancel the show?" Herol asked hopefully.

"Nah, it's okay — was thinkin' about you, me, and Marge. Everything's gonna be different tomorrow."

"What in hell you saying? I don't know nuthin' you're talkin' about all morning! — You ain't feelin' good. I'm gonna tell Happy Bottom!"

"You do and I'll flatten your head. Pull them buckskins back on — I gotta apologize to Pop."

Tyson's wrath submerged as Marge stepped through the doorway again, carrying two hot sandwiches on paper plates.

"I know you boys are anxious to get started, but have a warm snack before you go out in the cold." She handed them each a plate.

"Thanks!" Bob said.

Herol sat on the bench next to the file cabinet flexing his arms and shoulders as he tried to get comfortable wearing the tight harness under his buckskins.

Tyson offered Carter's chair to Marge then sat down on the desk beside her.

"Bet you made these, they're good," Tyson smiled.

"Will you be ready soon, gentlemen?" HB's pink, round face peered across the office at them from the doorway.

"Soon," Tyson answered flatly.

"Very well, I'll announce that you'll be departing shortly," he replied, and disappeared into the hangar.

Pop hunched his shoulders against the sliding hangar door and pushed until the frozen iron casters squealed. The wooden door creaked in protest as it jolted over the half-buried tracks. It slowed to a stop in about ten feet, ample room for the impatient guests that crowded through with their folding chairs, intent on having a grandstand seat for the show.

Carter and Tyson sat in the front cockpit of their planes, crouching low to avoid the snow avalanche that churned beneath the whirling propellers and corkscrewed over the wings and into the cockpit.

Marge sat near the center of the row next to the vacant chair reserved for HB, who was busy providing seats for those who were standing. While the others chattered with excitement, her interest was focused on the flightline activity. Herol was barely visible through the blowing snow as he stood on the wingwalk of the Eaglerock talking to Carter. A parachute, slung over his shoulders, beat a tatoo on his back as he crouched against the propeller blast. Finally, he stepped off the wing into the snow and trudged slowly toward the Standard.

Pop stood on the wing talking to Tyson, then he waved farewell and made his way toward the Eaglerock.

"Good luck, son," Pop greeted Herol as they approached.

"Thanks!"

"I'll check the skis before you start — they're probably frozen again."

"Okay."

He pulled off his parachute and strapped it into the rear cockpit of the Standard. Tyson turned around, offering his mittened hand, and helped him onto the wingwalk.

"I'm ready to go whenever you are," he shouted over the engine.

"I'm ready!" Herol yelled.

"We'll make it short and sweet. I'll pick you up on the ladder from the rear — don't leave the cockpit! Let me fly into position over you. I'll be watching when you make contact and will pull away. After the pickup, we'll make a low pass, then you climb aboard and I'll make a jump. Land and pick me up. Good luck!"

"Thanks! Pop's on his way over to pry the skis loose — good luck!"

Tyson gunned the Standard into a wide turn and coasted onto the hangar ramp, the Hisso idling slowly while he brushed the snow from his goggles with a leather mitten. He waved a farewell to Marge, then pushed the throttle ahead and the propeller churned up another blizzard and when the snow settled once again . . . he was gone.

Carter followed in the ski tracks of the Standard and stopped momentarily in front of the crowd. He nodded to his audience, and Herol waved from the rear cockpit. Moments later they were hightailing across the snow.

Tyson peered over his left shoulder to catch a glimpse of the Eaglerock about a mile behind and climbing fast to intercept him. He wheeled the Standard on her left wing tip and slipped into position, above and to the rear as the Eaglerock leveled off.

They flew a tight formation and circled back to the field. Carter decreased speed to a slow cruise, giving Tyson reserve engine power to maneuver while making the ladder contact.

As they flew low across the field, Tyson slid the skis of the Standard to within inches of the Eaglerock's top wing. He sighted through the Hisso's propeller arc in line with the Eaglerock's and synchronized his engine speed by matching the shadow forms of the two propellers. Now he had a reference power setting for the precise flying he must do.

Herol sat facing to the rear in the Eaglerock, mindful of what Tyson was doing. They would soon be ready for the first act.

The Standard pulled up and resumed its former position while Carter led them in a gentle turn across the field to a position that would place them in view of the spectators.

They descended slowly to a hundred feet above the snow. Carter leveled off and held a steady course while Tyson backed off and released the ladder. It uncoiled and dropped from the ski and snapped taut below the lower left wing as the safety cables near the cockpit took hold.

Herol examined the steel hooks protruding through his gloves. Then he checked for loose clothing and rose to a crouching position in the cockpit as Tyson maneuvered the ladder toward him. The bottom rung cleared the Eaglerock's tail by a hair and swung closer.

The Standard hovered overhead, inching down on Herol's signals, while the ladder lashed back and forth in the propeller wash.

He stood up, straining to catch the second or third rung when the ladder danced within his reach. He lunged at it with both hands, and the steel hook in his right glove made solid contact while his left arm relaxed and his fingers let go. Instinctively, he pulled his knees toward his chest, reducing the strain on his right arm and supporting his weight in the canvas saddle.

The Standard roared up and away, lifting him out of the cockpit and scant inches clear of the Eaglerock's upper wing. He held his breath until his heart stopped pounding, then strained for the ladder with his left hand and engaged the hook — safe for the time being.

He looked between his legs at the cheering crowd below, then above him at the Standard and Tyson's grin of approval.

Carter now let Tyson assume the lead as he tucked the Eaglerock's wingtip next to the rear cockpit of the Standard. Act two was coming up!

They roared past the hangar ramp in formation, waving at the spectators, while Herol climbed onto the ladder and rested in his precarious saddle. They sped downwind in a climbing turn, and as they reached the end of the field Carter peeled away in a graceful Split-S to a landing.

The Standard clawed for altitude while Tyson held it in a gentle bank to bring them in line with the hangar. In the meantime, Herol was making careful progress up the ladder.

When he reached the top, he stepped onto the heel of the ski. He was out of breath and the ski platform felt strange, but it was a needed rest stop.

Tyson held the plane steady and retarded the throttle, giving Herol a temporary respite from the propeller blast.

"Good show, old buddy! Come aboard when you get rested," Tyson shouted over the engine.

"I'm froze stiff! I'm comin' now!"

The propeller unwound to a fast idle and he scrambled onto the lower wing. Tyson reached outside the cockpit and pulled a release pin and the ladder trailed away to the field below.

"I'm cold!" Herol panted, trying to catch his breath.

Tyson reached out of the cockpit and held a protective arm around his waist while he held snug against a center section strut.

"Get your wind back while I make another pass across the field. "I'll make the jump as soon as you're ready." He edged the throttle forward and leaned into a slow turn while Herol climbed into the rear cockpit.

The biplane seemed to hang motionless over the center of the field as it labored against the choppy air. Nervous winds scooped fresh snow from the ground and battered it down again. Dark rolling clouds in the northwest were banding together and spilling thin columns of snow across the open fields. The backside of the morning's blizzard would soon lay another white blanket across the land.

Herol was flying the plane now and he tried to coax a little more lift out of the straining wings, while Tyson busied himself with a last-minute check of his parachute. Time was running out for them. Already snow flurries were sifting out of the overcast as they approached the base of the storm.

He thumped on the cowling with his fist and slowed the engine down to a rumble. "I can only get about fifteen hundred feet — wanna cancel the jump?"

"It's okay. Turn downwind and get in position. I'll be ready!" Tyson yelled.

They skirted around snow showers, hanging close to the overcast and riding the gusty air downwind past the end of the field. A ragged turn in the turbulent skies brought them back in line with the hangar.

Tyson was anxious to get the show over with. He glanced at his watch — only three hours before his wedding! It was time to wrap it up! He hoisted himself to a standing position in the front cockpit, both mittened hands grasping firmly to the center section struts. The frigid blast from the propeller startled him at first. It was a different world than his years of summertime performances. He could only guess how the kid's exposure on the ladder must have felt.

Herol fought the ungainly Standard to keep her on an even keel. Sharp gusts wrinkled the fabric wings, making the interplane wires dance, and sudden downdrafts kept the sputtering Hisso gasping for fuel. The edge of the field passed beneath their wings and Tyson glanced over his shoulder and nodded. They'd soon be over midfield, and he'd play the finale of their last performance together.

He lifted his foot over the edge of the cockpit and planted it firmly down on the lower wing, then debated as to whether he should bail out from here, or juice it up a bit and make a "pull-off" from the center bay of struts. The turbulence presented a problem. First he'd get out, and then he could play it by ear, as he'd always done.

Herol concentrated on holding the plane as steady as possible when he saw Tyson lifting his other leg out of the cockpit. Suddenly, the plane cartwheeled off on its right wing as it was buffeted by a violent downdraft, catapulting Tyson against the underside of the top wing and just as quickly dumping him headfirst back into the cockpit.

"You hurt, Bob? You okay?" Herol shouted against the wind.

Moments later, Tyson's head appeared above the cockpit. He was looking to the rear with a dazed expression. He was seriously injured. Blood poured through a shattered lens of his goggles and streamed down his cheek. His nose was broken and dripped blood down his chin. He tried to grin and braced himself to climb out again.

"Stay down! Don't get up!" Herol screamed.

His parachute blossomed, then swirled past him, sucking yards of silk into the rear cockpit. Dozens of lanyards held fast to a shredded panel that whipped against the side of the fuselage and lodged in the tail. The chute was ensnarled under the front seat, giving them a temporary respite for the time being.

Tyson stood up and reached into the rear cockpit, groping through silk to find the throttle, and when he did, pulled it to idle.

"You gotta bail out or we'll both be killed!" Tyson hollered.

"Like hell! Pull off your harness and I'll try to get us down!"

"Not enough control! — I gotta tear loose and take my chances. Do a half roll. You bail out and I'll get loose at the same time."

"I'm stickin' with you!" Herol screamed.

"I'm doin' the thinkin'! Unbuckle your safety belt, roll us over, and we'll both go! — Do it!"

Carter taxied the Eaglerock against a fresh quartering tailwind on his way to the hangar. Progress was slow as he peered blindly through the propeller blast that churned tornadic snow storms in all directions and blanketed the plane. Frequent capricious winds half buried the wing tips in the snow as he fought the rudder to keep from weathercocking into the wind.

He heard the Standard droning overhead and cocked his head toward the sound, catching a flash of the sun's mirror image atop the silver wings. Then he saw the parachute trailing outside the fuselage!

"Godamighty! They're in trouble!" he whispered.

He shut off the ignition and leaped out of the cockpit before the propeller unwound, stumbling through the snow until he reached the hangar.

Curious spectators looked his way as he raced past them unaware of the crisis that was in progress above them. He spoke briefly to HB, grabbed Marge by the arm, and rushed both of them toward the office.

He ran into Pop at the hangar door.

"Get the truck fired up and see if you can get out to midfield!"

"Yeah, I know — I'm on my way!"

Herol half-rolled the Standard and he and Tyson fell away. The two parachutes opened seconds apart beneath the empty cockpits. Tyson's weight ripped his chute loose from the plane and the canopy snapped open immediately, almost entangling him again in the tailplane. Herol fell free past Tyson and was about fifty feet below him.

The empty Standard wandered through an aimless turn, then inclined its nose in a descending arc and dove inverted to the ground. It hit with a thud near the upwind end of the field and burst into an inferno. A gyser of oily smoke shot straight up, then leaned with the wind and carried yards of burning fabric down the field.

Shredded silk fluttered from a torn panel on Tyson's chute, threatening to collapse the canopy as he swayed violently in the rising smoke clouds from the burning plane. He'd be faced with a hard drop, but the blanket of snowcover of the field might help cushion the fall.

Herol tugged at the risers on his chute trying to spill more air from beneath the canopy so that he'd match Tyson's rapid descent. The brisk wind was drifting them far down the field and further away from the immediate aid that Tyson needed. He could see him hanging limp in his harness and making no attempt to control his chute. Herol guessed that he might be in shock from his injuries.

A clap of thunder erupted from the storm overhead and spilled out gale force winds in its wake. Tyson was near the ground and drifting rapidly toward a cattle fence on the end of the field and a stand of pine trees on the side. Herol was close behind, slipping his chute as much as he dared, intent on landing near his injured partner.

A viscious gust hit them broadside, swinging their chutes like pendulums. Tyson's canopy exploded as the torn panel ripped wide open. The chute collapsed, trailing a long streamer behind, and he plummeted to the ground.

Herol tumbled end over end when he hit, the gale winds dragging him across the snow until he plowed into the fence. He squirmed out of his harness and stumbled toward his partner, who lay unconscious near the trunk of a tree.

In the meantime, Pop was driving the Model A as fast as it could go. The truck was loaded with blankets, a stretcher, and first aid supplies. HB had volunteered to drive Marge and Carter to the scene and was following close behind Pop's truck. A few curious guests started out on foot but soon gave up and returned to the warmth of the hangar. An ambulance would eventually arrive if it could get through the falling snow.

Tyson lay sprawled on his back in deep snow, about an arm's length from a tree. It was obvious that he'd drifted downwind nearly as fast as he fell. His body had ploughed a wide furrow in the snow from upwind, about a hundred feet to where he lay. He'd lost his goggles sometime during the descent, exposing his battered features. His injured eye was buried in a pool of congealed blood, and the other was swollen shut. His nose was badly misshapen and dribbling blood into his mouth. His lungs rattled as he labored to breath.

Herol knelt down beside Tyson, unsure what he should

do next. He couldn't believe that anything like this would ever happen to them. He shielded Tyson's head with his body from the drifting snow while he pulled down the parachute that was lodged in the branches overhead; then he fashioned a pillow from the silk and placed it under his head and made a blanket with the remainder and tucked it around him.

The Model A slid to a stop. Pop stepped out laden with first aid supplies and hurried toward them.

"Get the blankets and stretcher! I'll see what I can do."

He reached inside Tyson's collar with his fingers and felt for a pulse in his neck. Satisfied that he was alive, he waved a bottle of smelling salts under his nose. Then realized that Tyson could breathe only through his mouth. As an alternative, he uncorked his bottle of moonshine and washed some of the dried blood from Tyson's face, then placed a few drops of alcohol in his mouth.

Tyson moved ever so little.

"What happened?" he moaned.

"Lay still! This is Pop — you're hurt. We're gonna get you fixed up."

"Is Herol all right?" Tyson mumbled.

"I'm okay — do like Pop says," Herol assured him.

HB's limousine pulled alongside the truck. The rear door swung open and Carter was out with Marge close behind, HB decided to confine himself to the heated comfort of his car for the time being.

"MARGE?" Tyson called out.

"I'm here," she said kneeling next to him.

"I can't see, but I knew it was you."

"I'll stay with you." She placed her hand in his.

Carter began scooping the snow from underneath Tyson and carefully feeling his arms and legs for broken bones.

"Somebody put blankets on the stretcher and be ready to cover him up!" Carter hollered.

"Pop? I'm hurtin' all over — you got a shot of booze or something?"

"Sure, here you are," he held the bottle to Tyson's mouth.

"I'm busted up inside!" He coughed out a mouthful of blood and fainted again.

"Let's get him on the stretcher now! Pop! Herol!"

They lifted their burden carefully onto the stretcher, covered him with woolen blankets, and hoisted him over the tailgate of the truck.

Herol climbed in and sat down beside Tyson while Carter helped Marge in. Pop hurried to get in the truck and started the engine.

"We'll follow close behind if you need me," Carter said as he stepped into HB's limousine.

"Marge? — Herol?" Tyson reached blindly for them.

"We're right here with you, Bob," she said, and wiped his lips with a piece of gauze.

They rode in silence through the soft snow, Marge fighting back tears as she cushioned Tyson's head in her arms. Herol kept a firm grip on the stretcher and stared unseeing at the burned-out wreckage near the end of the field.

"Why, Marge? Why did it happen? It should have been me! He was always chewing me out for doing dumb things!"

"He'd want it this way," she said softly.

"That's right, Marge. It was our last encore, a show they'll never forget!" Tyson was smiling as best he could.

"Oh Bob!" was all that she could say between tears.

"It's the last sunset," Tyson went on. "I been counting them like Johnnie said." He turned his head toward Marge and seemed to relax.

"Don't listen to him. Them is goofy things he heard a long time ago," Herol said.

"No, *you* heard Johnnie Day say that — *he* knew! I was with him!" she whispered.

"Aw! Forget it! We gotta get Bob fixed up." Herol tried to smile.

Pop slowed down as they approached the hangar ramp, and HB parked his limousine in front of the hangar doors.

"Let's try to get these people back to town, and do something with the Standard," HB said as he opened the car door.

"Don't worry — I'll take care of it right away. Pop and I will salvage what we can from the Standard and sell the rest for junk," Carter replied.

"Very well. I'll be with Tyson," then he hurried away.

HB approached them, searching for words that might be of comfort. Herol was sitting motionless, staring at his partner. Marge was crying while she cradled Tyson's head in her arms.

"You're all right, Herol?" HB asked.

"Guess so. Just Bob — he's hurtin'."

"Bob, this is HB."

"Hi!"

"I blame only myself for what happened today. It was too late in the year for a show. I should have never asked you."

"That's okay, just bad luck," Tyson said softly.

"You and Marge will be looked after, and I'll take care of all expenses so that you'll be well again soon."

"No deal! Unless you include my partner too!" Tyson's harsh concern flared momentarily.

"Absolutely! Herol too!" HB smiled.

"Thanks! We ain't got much time before the wedding — gotta hurry!" Tyson reached for her hand.

"Oh! Dear God!" Marge wept and held him close to her.

"Maybe we can do it at the hospital, or somethin'," Bob's voice trailed away to silence.

Tyson turned his head toward Herol. "Everything's gonna be different tomorrow. You still don't know much about living, but you sure as hell will before long. It's all pretty simple. You're born and you die — and in between is just a second of eternity.

Marge held him a bit closer and continued to cry. Herol stared at nothing. HB felt utterly hopeless.

"He's talking out of his head again," Herol glanced at Marge.

"Let the dear man rest!" she sobbed.

The ambulance siren cried its loud alarm across the open field as it skidded through a turn on the road leading to the main hangar. It rolled to a stop beside the truck, and two attendants jumped out and raced to the rear. One grabbed the door and swung it open, then they pulled out a lightweight portable bed and set it on the ground.

"Better let the ambulance crew take over," HB said, and stepped out of the way.

"We'll move the patient. Please stand aside!" an attendant said.

The other one climbed into the truck and hurriedly examined Tyson's injuries. Then the two of them lowered him to the ground and, crouching on either side, they gently lifted him off the stretcher onto the portable bed and slid him into the rear of the ambulance.

"May I go along?" Marge asked.

"Yes ma'm, but nobody else! Hurry!"

He helped her get in, then climbed aboard and slammed the door shut. The driver started the engine, the siren wailed, and they were gone.

"You can ride into town with me," HB said as he walked toward his car.

"Thanks," Herol said absently.

Herol watched the ambulance disappear and knew that half of his life was vanishing in front of him as they sped away.

"Herol?"

"Coming!"

It was all over! No goodbyes! He felt lightheaded and knew that he'd surely be sick.

"Everything's gonna be different tomorrow!" Bob's prophetic words were ringing true.

Herol was frightened, a sense of loneliness crept over him. The uncertain future held little promise other than a repetition of the past.

Why did it happen to Bob?

When would it end?

Would he be next?

He buried his face in his hands and wept.

Herol's great airplane and faithful companion, his Hispano-Suiza, 'Hisso', powered A-4 Eaglerock. (Author's collection)

17 EPILOGUE

Herol glanced over the edge of the cockpit and watched the tall chimneys of a pulp mill, near Regina, glide under his wing. The Eaglerock leaned to the left and settled down to an easterly heading that would parallel the iron tracks below and guide them eventually to St. Boniface, about a third of the distance to Ottawa.

The arctic skies were an iridescent blue canopy that seemed to extend beyond eternity. To the north and the west the vivid blue melted into the bright snowscape in the distance. In front of him the horizon made a crisp blue-white line across the engine cowling of the Eaglerock. The big Hisso purred like a tomcat as they raced in front of a stiff tailwind, hightailing across the miles of snowbound wilderness.

He spotted a deer trail below, along the edge of a clearing, and was tempted to touch his skis in the snow "one last time" and leave his own tracks alongside, but changed his mind. Bob would call him a 'kid' for sure if he saw him horsing around.

The Eaglerock lazied into a gentle turn and they got back on course. There were hundreds of miles of wilderness stretching ahead, plus the long trek back to Bemidji. He decided that if he was going to live up to Bob's expectations, it would have to be all business from now on — that's the way Bob would have wanted it.

He looked at the cowled-over front cockpit and was reminded that the honeymoon couple should be there instead of the reserve 40-gallon fuel tank. Maybe Bob planned it that way! — It couldn't be!

He smiled as he thought of his dumb 'first solo' in Coeur d'Alene, when he and Bernie got in trouble with the parachute jump — and Bob's surprise reaction, only to be reminded that from now on it was all for real. It would be a 'first solo' for the rest of his life because he knew now that he was completely alone.

He recalled the last words of advice that Gramps offered him eons ago.

"The Heavens belong to the Almighty, just as the sea is part of His realm. He might welcome you, or He might resent your intrusion. It's up to you to decide if the rewards you hope to gain are worth the price you might have to pay."

"I gotta try," Herol said to himself.

The author stands proudly in front of the 1927 Waco 10 that he restored.

About the author . . .

Author Harold Salut was born in Bemidji, Minnesota. As a boy he often skipped school to wash airplanes and do odd jobs around the airport in Fargo, North Dakota. The joy of being a part of the flyer's world far exceeded his worries over threats of being expelled from school.

He took his first flying lesson in 1931 and made his first solo flight in 1932. Shortly after that he began his flying career, barnstorming the northwestern United States and Canada as a stuntman and parachute jumper.

Mr. Salut served in the U.S. Navy during the war years as a Naval Aviation Pilot. At the end of the war he taught seaplane and instrument flying at Hasbrouck Heights, New Jersey. In the years after the war he also participated in air shows and air racing competition.

Harold rounded out his aviation career by flying for the Civil Aeronautics Administration as a flight inspection

Left: The author during the 1940's at the controls of a multi-engined transport. Right: The author during the 1960's beside a Lockheed P-80 Shooting Star. One of the many jets he has flown during his career.

pilot. He flew single and multi-engined jets for the last thirteen years of his career.

On retirement in 1973, Mr. Salut reverted to "amateur pilot" status after having a flying career that spanned thirty thousand flying hours from the "sticks, baling wire and rags" era to the days of modern jet transports.

In 1972 Mr. Salut was awarded an honorary Doctor of Humane Letters by the Institute of Applied Research, London, England.

Mr. Salut is an active member in the following:

- National Aviation Hall of Fame (Charter Member)
- OX5 Pioneer Airmen of America
- Experimental Aircraft Association
- Antique Aircraft Association
- Long Island Early Flyers Club
- S.E.A. Silver Eagles Association (U.S. Navy enlisted pilots)
- Tail Hook Association (U.S. Navy aircraft carrier pilots)
- Silver Wings — Aviation Pathfinders